THE
NORM
CHRONICLES

Despite starting out telling stories as an English Literature graduate working in journalism, **MICHAEL BLASTLAND** somehow learned to count, devising the Radio 4 programme about numbers *More or Less* and writing, with Andrew Dilnot, *The Tiger That Isn't*, a guide to numbers in the news. On ice skates, he is life-threatening (mostly it's his own life) but more afraid of other people's raised umbrellas. His other risk dislikes include confined spaces, heights and fairground rides.

DAVID SPIEGELHALTER is a statistician who rejoices in the title of Winton Professor for the Public Understanding of Risk at Cambridge University. He is, or was, a proper academic and has far too many letters after his name, but feels his greatest achievement is not doing badly in the risky TV programme *Winter Wipeout*. He lives in a flood zone, but is more anxious about forgetting where he put the house keys. He particularly likes heights, confined spaces and fairground rides.

Together, Blastland and Spiegelhalter sound like a Dickensian music-hall act or a firm of dodgy solicitors. But they think that combining ideas and perspectives is the best way to make sense of the subjects of this book: danger, risk and chance – subjects that could be said to exist, as they hope to show, only in a clash of viewpoints.

For Ron and Shirley – MB

For Kate, Kate and Rosie, for putting up with me – DS

THE NORM CHRONICLES

Stories and Numbers about Danger

MICHAEL BLASTLAND
and
DAVID SPIEGELHALTER

PROFILE BOOKS

First published in Great Britain in 2013 by
PROFILE BOOKS LTD
3A Exmouth House
Pine Street
London EC1R 0JH
www.profilebooks.com

Copyright © Michael Blastland and David Spiegelhalter, 2013

3 5 7 9 10 8 6 4 2

Typeset in Garamond by MacGuru Ltd
info@macguru.org.uk

Printed and bound in Great Britain by
Clays, Bungay, Suffolk

The moral right of the authors has been asserted.

A CIP catalogue record for this book is available from the British Library.

ISBN 978 1 84668 620 7
eISBN 978 1 84765 829 6

The paper this book is printed on is certified by the © 1996 Forest Stewardship
Council A.C. (FSC). It is ancient-forest friendly. The printer holds FSC chain of
custody SGS-COC-2061

CONTENTS

INTRODUCTION

A SHORT STORY about danger:
One day, by devilish coincidence, three people travelling separately on the tube – Norm, Prudence and Kelvin – saw three unattended bags.

For Norm it was a dusty-blue canvas hold-all on the floor, tucked against the seat on his side. At first, he thought nothing of it. Then he looked again, and then up and down the carriage. Almost empty.

'Calm, Norm,' he said to himself, and reached down to pull up his socks – the green thermal ones – glancing sideways for wires. He sat up and forced himself to focus on the probabilities, scratched his nose, concluded several times that the bag had been forgotten, that's all, stood up and went slowly to the far doors to get off at the next stop and enjoy the extra walk.

When Prudence looked up from *Fifty Shades of Grey* to see a pristine rucksack on the seat opposite, she felt quickly sick. If it has a travel tag, someone is taking it somewhere. If it doesn't ...

It didn't. She thought of her children, motherless, crying, and lost the strength to move. Her mind filled with hellish images of herself blown apart and her hair ruined.

Counting out the last seconds of her life, ticking to the blast, she gestured and mouthed at a passenger standing nearby, a warning, a hope: 'a ... bag,' she mumbled, and pointed like the ghost she was becoming.

'Oh, yeah,' he said, and grabbed it. 'Cheers.'

And Kelvin? When he clocked a black briefcase as the doors slid open, he opened it – what else? Picked it up, sat down, flipped up the sucker's lid, took out a folded *Daily Telegraph* and slid it into the side-pocket of his leather jacket, shuftied a wad of paperwork, noticed a foil wrap down the side, teased it open, lifted it to his nose, snorted – with an eye on the teenage girl further down, doing her lashes – threw in the wrapper, closed the lid, put down the bag, sat back and closed his eyes.

Three people, three points of view about danger, to which we could add many more. What's yours? Danger, as every experience and a million stories tell us, is in the nerves of the beholder.

But not entirely. There are also numbers.

Here are two. The first is terrible and well known: on 7 July 2005, 52 people died in terrorist bomb attacks on underground trains and a bus in central London. Second, in 2011 about 30,000 bags were left on London Transport. Let that sink in: 30,000.

So, is an unattended bag on the tube dangerous? How do the numbers and stories compare: the particular stories of the 52, and those of Norm, Kelvin and Prudence?

Leave that question burning for a moment, for another short story, a true and famous one.

One day Anna and her friends went skiing. Anna was a good skier, and in any case there's not much chance that skiing will kill you. Then she lost control. She fell onto her back on a frozen stream near a water-fall. A hole opened in the ice. Freezing water poured into her clothes and pulled her in, head first.

She would have been dead in minutes, but under the ice she found an air pocket. She could breathe, for a while. First her friends, then a rescue team, tried to pull her out. They failed. They tried to dig her out. The ice was too hard.

Anna remained conscious for 40 minutes. But finally her breathing slowed, and then stopped. Then so did her pulse. It was another 40 minutes before she was pulled clear.

Normal body temperature is 37°Celsius. Hypothermia begins to set in at about 35°C. When Anna arrived at hospital, her temperature was 13.7°C. No one that cold had ever lived.

But the doctors did not give up. Slowly, patiently, they warmed her blood outside her body, then pumped it back into her veins. More than three hours after she stopped breathing, more than two hours after she arrived at hospital, Anna's heart began to beat again.

But when she woke, ten days later, it was to find herself paralysed from the neck down, and one of her first feelings was anger that she had been revived, for this. Then she made an almost full recovery. A few years later she was working as a radiologist at the hospital that saved her life. She still goes skiing.*

Anna's story is now celebrated – as a marvel of survival and a medical revelation. But for us it makes a different point that has little to do with human endurance or scientific understanding of the effects of extreme cold, and is simply this: Anna rode the mother of all fortune's roller-coasters. At every turn in a twisting tale, at every roll of the dice for good or bad, it was as if she threw six 6s.† All her luck was extreme.

Anyone can fall, even those who ski well. But how and where Anna fell – into that concurrence of water, hole and hard ice – was absurdly unlikely. Then to find an air pocket was a godsend, or seemed so, but for the maddening, desperate twist that made it so hard to pull her out. To be visibly dead from such extreme cold and yet live was impossible, until she did it; then to survive but in the end wake up paralysed, except that the end was still to come with almost full recovery, was one amazement after another. And behind every turn was the most twisted of all: that it was near-death that saved her, when fatally bitter cold turned out – by slowing her metabolism down nearly to a stand-still at just the right moment – to preserve a kernel of life precisely when breath stopped. Life is sometimes improbable.

What our brief sample of stories and numbers tells us is that risk

---

* The story of Anna Bagenholm comes from various sources, including *The Lancet*[1] and Atul Gawande, in his book *Better*.[2]

† The usual phrase here is that Anna beat or defied the odds. Strictly speaking, no one can defy the odds. Odds simply describe how many people are expected to be on each side of a possibility. Even a million to one against when you turn out to be the one is not beating the odds: it *is* the odds.

is two-faced: on one side are the seemingly hard-nosed calculations of probability, such as the 20 per cent extra risk of cancer from eating sausages that hits the headlines, or the infinitesimal percentage of bags, unattended or otherwise, that explode in London, or the chance of surviving if your body freezes, you stop breathing and your heart quits; on the other side are people and their stories, like Anna, or the 52.

Numbers and probabilities tend to show the final account, the risks to humans en masse, chance in aggregate summarised for whole populations. These numbers reveal hypnotic patterns and rich information. But they are indifferent to fate and its drama. Numbers can't care and don't care; life and death are percentages, unafraid of danger, shrugging at survival, stating only what's risky, what's not, or to what degree, on average. They are silent about how much any of this, right down to a love or fear of sausages or ski slopes, matters.

But we – and you – are not averages. We are also subjective, we do care and might even argue about skiing, terrorism and sausages. We have our instincts, feelings, hopes, fears and confusions. Our intuition might not match the stats, and we might say, 'So what, I'm out of here.' Or maybe we see danger and take the leap anyway, base-jumping from a cliff edge in a wing-suit because we love the buzz (see extreme sports, in Chapter 16); maybe we run screaming from spiders (phobias appear in Chapter 25). We ask, 'Will I be safe? Will my children be safe?' But also 'Will I be in control?' (Chapter 15), 'Will I be happy?', 'Do I want this thrill?', 'Do I value this choice?' (drugs are in Chapter 9), 'Should I take this chance?' And 'How does it feel, and what's it worth' (see childbirth, in Chapter 11)?

An extreme illustration of the difference is that, as far as the mortality statistics go – the statistics used to calculate the risk of skiing, along with many of life's other hazards – nothing happened to Anna. She was a tick, not a cross; she lived and didn't die, and that's it, all that the mortality record has to say.

Danger is the shark in shallow waters, the pills in the cupboard or a grand piano teetering on a window ledge while children skip below. It is the diet too rich in cream, the base-jump, the booze, the pedestrian and the double-decker, driving a car fast or the threat of weirder weather. It is the spills and the thrills. In other words, danger is everywhere and always. And in all cases we find those same two faces: one impassive, formal, calculating, the other full of human hopes and fears.

The unusual aim of this book is to see both at once. We hope to show people and their stories *and* the numbers, together. We set out to do this mainly to explore how these two perspectives compare, but along the way we found that this raised an awkward question: are the two faces of risk compatible? Can risk claim to be true to the numbers and to you at the same time? We will present both sides as we try to find out, but we will tell you our conclusion now.

It can't. For people, probability doesn't exist.

That's an extraordinary claim from writers sometimes geeky enough to have two hoods on their anoraks. But with a little luck, the proof of it – and exactly what it means – will emerge through the clash of perspectives in this book.

The numbers and probabilities are all here. With them we show the chances of a variety of life's tricks and traps: risks to children; risks of violence, accident and crime; dangers from sex, drugs, travel, diet, lifestyle; risks of natural disaster and more. We say how we know these risks, why sometimes we can't know them and how they've changed, and we use the best methods we can find or invent to make them easy to grasp. In particular, we use a cunning little device called the MicroMort and a new one called the MicroLife, two friendly units of deadly risk that we think offer real insight. You will meet them soon enough. In this respect, the book is a new guide to life's odds.

The human factor is here too. People don't always do what the numbers seem to suggest they should. Some feel safe when they are in danger and in danger when they are safe, while the numbers may matter less to us than feelings of power or freedom, our values, our likes and dislikes and our emotions.

One reaction to the difference is to tell people they are stupid, and that if only they listened to the experts they'd live longer and sleep sounder. Another is to say that the experts may be right about the averages but that they clearly never had kids or an undiagnosed chest pain, or wanted to take a corner too fast.

Either way, the human factor can't be ignored. To show it, we use a technique that is, well, risky: we combine fact with fiction, numbers with stories. Why write a book that's part numbers and part stories? Because that is how people see risk – through both stories and numbers.

Each has its virtues and shortcomings. Numbers tell us the odds. Stories are how we often convey the feelings and values that numbers cannot, feelings and values that might in turn distort our perception of the odds. Stories impose order, but often artificially – beginnings, middles and ends, all tied neatly together (too neatly?) with cause and effect. Numbers give us probabilities, which often don't claim to know the precise causes and effects of how one thing leads to another but simply show us how it all adds up into a tally of life and death. To understand how these perspectives play out, shouldn't we see them together? To be true to them, shouldn't we try to let each speak on its own terms?*

Steven Pinker wrote in *The Blank Slate*:

> Fictional narratives supply us with a mental catalogue of the fatal conundrums we might face someday and the outcomes of strategies we could deploy in them. What are the options if I were to suspect that my uncle killed my father, took his position, and married my mother? If my hapless older brother got no respect in the family, are there circumstances that might lead him to betray me? [...] What's the worst that could happen if I had an affair to spice up my boring life as the wife of a country doctor? [...] The answers are to be found in any bookstore or any video store.

So we created some characters. First, Norm, the one who saw the blue

---

* What is a story? David Herman, editor of the *Routledge Encyclopaedia of Narrative Theory*, says: 'Stories are accounts of what happened to particular people and of what it was like for them to experience what happened.' That'll do for us. It's the 'particular people' bit that matters, which includes both fact and fiction. Specialists use the word 'narrative' to distinguish the way a story is told from the pure events, but this is not a book about narrative theory, although here and there we do talk about the story form – heroic medical stories in Chapter 23, for example.

hold-all on the tube and tried to compute the optimal, proportionate reaction: our hero. Something or someone is out to get Norm, though he's done nothing wrong. He's just an average guy (the clue is in the name), looking for a safe path through life. He's so average that even his attempts to stand out are average, the sort of guy who feels moderately about Marmite.

But life has its own plans for Norm: maybe a car crash, a fatal dose of bird flu, a mugger's knife or meteorite, a nuclear meltdown or his own spreading waistline. Somewhere, an assassin waits.

Still, he tries. Risk is calculable, he says, and with reason as his guide and a sense of proportion he can steer a true course. His habits are ordinary, he likes a nice cup of tea but not too many, wears M&S trousers and invites little risk from hot passion or daring. Even so, someone or something wants him dead. Norm's entire, blameless life is a story of mortal danger, as to some extent is yours, and ours.

Prudence (another clue), the one who panicked on the tube, treads warily, all anxious glances. Every stranger's footstep could be following her. The numbers hardly matter, and one scary story is all that fear needs to set fire to her imagination.*

Finally, there are the Kevlin brothers, Kelvin, Kevin and Kieran, chancers and risk-junkies who fly by the seat of their pants and might, just for the hell of it, tell you exactly where to shove your reasons and probabilities.

Side by side, chapter by chapter, numbers and stories. We had planned to let the clash of perspectives stand on its own without comment. In the end we went a step further to bring out the differences. So within the non-fiction we also explore the psychology of risk perception. This is where numbers and stories meet, and often disagree.

All of which leaves us with two competing world-views. What happens when they collide? A fight, often as not, in which advocates of one approach accuse the others of being irrational and the others reply that the first lot are unfeeling.

These conflicts are elemental. Within people's attitudes to risk lurk

---

* There's a tendency for women to be more risk-averse than men, but only on average. We are aware that we play with stereotypes all round.

many of life's deepest tensions. Pick your side: art versus science, feeling versus reason, words versus numbers, perception versus objectivity, stories versus stats, instinct versus analysis, the particular versus the abstract, romanticism versus classicism, red trainers versus brown Hush Puppies: in short, the eternal row between fundamentally different versions of truth and experience. It is easy to set up camp in one or the other and never look outside.

Even if you think you don't take these sides, and certainly not so crudely, you might take them over danger. It touches something deep in our attitude to life; it helps define what kind of people we are. Sometimes we embody both sides and are torn by contrary impulses as the struggles between different world-views – sometimes the numbers, sometimes the stories and emotions – ebb and flow as we try to work out how to live. Beside such weighty stuff, the fact that danger might also strangle us with a blind cord, poison us with salmonella or blow us to bits seems almost by the way.

Here's a quick illustration.

It's a summer's day. You are walking down the High Street licking a vanilla Cornetto when a red number 42 double-decker bus thumps onto the pavement, whacks your ice cream from your hand and rips into Tesco Metro in a blizzard of glass and twisted metal, leaving you shocked – but unharmed. What are the chances?

In no instance are the exact facts of an accident predictable with anything like certainty. True, we know, to take another example, that leaning out from a ladder to paint the irritating bit in the far corner is asking for it. We can see what's coming, we say. But can we? Reliably? Of course not. Sometimes it comes, and sometimes it doesn't. Chance always plays a part.

Similarly, a bus crash might be mechanical failure or driver error; it might depend on the timetable, the road, the traffic, on the weather, on you, on the length of the queue for the ice cream, which in turn depends on everyone in it, or it might depend on every one of those things – in all, an infinitely intricate and improbable spaghetti of causes, events and people. You, reading this *word* in this book at whatever time and place you happen to be, took an absurd quantity of cause and effect – right back to the beginning of everything, if you must. Which is another way of saying that no one knows the future. Life is too complicated.

And yet you might not be all that surprised if a crash like that really happened – if not to you, then to someone. We know for sure that countless things – unlikely or not – will happen somewhere to someone, as they must. More than that, we know that they will often happen in strange and predictable patterns. Fatal falls from ladders among the approximately 21 million men in England and Wales in the five years to 2010 were uncannily consistent, numbering 42, 54, 56, 53 and 47. For all the chance particulars that apply to any individual among 21 million individuals, the numbers are amazingly, fiendishly stable – unlike the ladders.[3] Some calculating God, painting fate by numbers up in the clouds, orders another splash of red: 'Hey, you in the dungarees, we're short this month.'

We know there will be accidents and incidents, and we often know what kind and how many – to the extent that we can predict pretty well how many people will be murdered in London by 28 July next year, and even how many murders there are likely to be on one day (which we did – in Chapter 22, on crime). Up close, life can appear chaotic. Every murder is unique and unpredictable, every fall and crash laced with infinite chance. Seen from above, people often move in patterns with spooky regularity.

This is the great puzzle about danger: that a million stories describe it, feelings inform it, and a million occasions conspire for or against every incident – and yet there are a relatively consistent number. Every cancer begins with a freak cellular mutation, and yet a fairly steady one-third of people will get one.* It is one of life's odder facts, this order amid disorder, the natural and spontaneous emergence of shape – persistent and predictable – even as everyone does their own thing.

So, from above, the course of human destiny is often clear. To

---

* Although, unlike falls from ladders, some numbers do change dramatically if some big casual factor has also changed, such as the decline in heart disease, largely from less smoking. Deaths from heart disease in men fell from 147 per 100,000 people in 2005 to about 108 per 100,000 in 2010, and for women from 69 to 48 per 100,000. These are huge changes.

individuals below, it is a maze of stories. It is as if there are two forces at the same time: one at the big scale pulling towards certainty, the other pushing individuals towards uncertainty. There's a word to describe this balance between the patterns of populations and the stumbling of a single soul, a word first used in its modern sense only a few hundred years ago: probability.

Probability – at least one version – begins with counting past events, such as '20 per cent of men who died in recent years, died from heart disease'. It then uses this to predict a pattern. 'About 20 per cent will die from heart disease in future.' But then it goes a step further: from that general prediction it gives odds for what will happen to individuals. 'The risk or chance that the average person will eventually die from heart disease is therefore also 20 per cent, or one in five for a man, about 14 per cent or one in seven for a woman.' Thus it moves from past to future, from the mass to the individual.

Probability is magical, a brilliant concept. It yokes together our two world-views, the two faces of risk: the orderly view of whole populations seen in numbers from above and the sometimes lonely view in the maze of stories below. It embraces all of us individuals in aggregated data. Today people use probability to help make decisions about everything from the weather to money or the chance of being burgled, the risks from mobile phones, sausages or tsunamis. It touches our hopes and fears at every turn. The news is full of it, and no wonder – it seems to offer a hold on the future. Which is why it is that little bit inconvenient that it doesn't exist.

Norm's life-course shapes the order of the risks here, as we discover how well numbers can guide him, from beginning to end. As for which to include, we chose whatever seemed interesting and personal. There's little, for example, about risks in business – or Enterprise Risk Management (ERM), as it is known – about which more than enough is said already.

To add to our ambitions, we hope that the different perspectives here create a book that everyone from all sides of the argument can read and enjoy on the path to mutual understanding. Although by trying to reach everyone, it has occurred to us that we might reach no one. It's a risk.

Finally, a couple of hazard warnings: first, people are learning more about risks all the time, so it won't be long before there'll be new data. This fact is not just awkward, it's relevant to the argument. What kind of trust can you place in numbers that don't stand still?

Second, in places this is almost a mini-encyclopaedia of hazards. There are a lot of numbers here, and they are more fun to dip into than read in one go. Even so, the statistical evidence could go on and on, and we welcome readers' arguments about all those extra stats that we left out. We had to stop somewhere.

So the scene is set. Stories will be pitched beside stats, reason will tangle with feeling and impulse, belief will quarrel with evidence. We looked hard at the data and threw them at an imagined life where objectivity doesn't always get a look in, let alone prevail, sometimes in spite of Norm's best efforts. In short, we have brought together as many oppositions around risk as we felt we could in one book and in the process hope to start – or restart – reflection on that mighty clash of worldviews. There's also something or other going on in a sub-plot about asteroids and the end of the world.

We've already stated our own conclusion about how all this ends. But what that conclusion means, whether the argument stands, how or whether the two sides can be reconciled, is ultimately, of course, for you to judge, as you also explore where you stand on danger, if you dare.

1

THE BEGINNING

H AD HE NOT POURED GIN in the fish tank in a stupid reflex
flick of the wrist when he tasted it – because the thing is, he
hated gin – the whole story would never even have begun.

It wouldn't have happened if the fish hadn't died, either. Or if he
hadn't felt bad for crashing the party in the first place and twigged that
the other girl knew who he was and might tell on him. But for all that, he
probably never would have been standing there the next day saying sorry.

'Erm, the fish ...,' he said.

'Yeah, the fish,' she said.

'Dead?'

'Yeah.'

'Right. I kinda knew that. I think it was me.'

'Uh huh.'

'So, erm, how much is a fish?'

'For one fish? ... dinner.'

'What? ... dinner?... Oh, OK! Yeah, dinner. But like, not like ... not
for every fish?'

'Hey, come on, fish killer!'

'OK, OK ...'

'Though to be honest ... they weren't my fish. But it's either dinner
or I tell my brother it was you and as he's a psychotic axe murderer. You
don't want that.'

'Right. Sure. So ... how many fish?'

'Forty-two.'

'Forty ...?!'

And had they not then discovered a shared love of Sudoku, sailing and an original sound recording of Alfred Tennyson's 'Charge of the Light Brigade', plus that he really liked her smile and she liked his hands and had a fascination for the strange birthmark on his right ear, well ...

'Incredible,' they often said afterwards.

'What are the odds against meeting like that?'

'But "forty-two"! It wasn't even true.'

'Exactly!'

So it was only after a heap of happenstance, a whole cocktail of accidental, it-could-all-so-easily-have-been-different events that they met again and talked and fell in love and had a baby – after he forgot the contraception the time they went camping but said 'What the hell' anyway – which really should have made the whole saga about a zillion to one against.

But then, when you think about it, everyone is improbable, everyone's story's a fluke. There are so many reasons why any one of us might not have happened. At least, every particular someone is improbable. There'll be people for sure, but why you?

As it was, by going back to say sorry, he was out when his flat caught fire and filled with suffocating fumes.

So as she lay screaming for an epidural and swearing that he was going to pay for this with his ass, was going to sleep with the fishes in fact, he was wondering about the baby's future, the strange course of luck and bad luck, the risks and coincidences of life, wondering how much in the riot of fates was calculable. What *are* the chances?

And at the very moment the baby was born, far away a spectacular fireball lit up the pre-dawn sky, a radiant explosion caused by the atmospheric entry of a small near-Earth asteroid just a few metres in diameter but weighing 80 tonnes and firing icily through 12 km of space per second, such that it shattered with the force of a thousand tonnes of dynamite and the brightness of a full moon into small meteorite fragments across the Nubian desert below.[1] The asteroid was named

Almahata Sitta. The baby was precisely 3,400 grams.* They named him ... Norm.

CAN NUMBERS HELP the infant Norm duck the slings and arrows of life? In *The Norm Chronicles* we will guide him with the best that we can find. We will also make them as clear as possible.

The last point – clarity – is a big one. There are lies and damned lies in risk statistics, for sure, but there's real information too, and a large part of the problem is cutting through to the good stuff and making it intelligible.

Say that Norm's dad is cooking sausages for the boy's tea when his ears prick up to a headline on the TV news that says eating an extra sausage – or is it a sausage every day? – something about a sausage anyway – increases our risk of cancer by 20 per cent. He pauses. Norm plays. The sausages sizzle.†

What does it mean, this percentage – 20 per cent more risky than what? Then he hears it referred to on the radio as a probability (and is that the same as a percentage?), and maybe later he'll read in the newspaper about the 'absolute risk' and a 'relative risk', and by now he's struggling, and who can blame him? But then there's another thing they talk about, called a 'risk ratio', and it all seems mathematical and maybe even rigorous – who knows? – and so some say 'you mean 20 per cent of people die from sausages?' and others 'you mean extra sausages cause 20 per cent of cancers?' or perhaps 'you mean 20 per cent of people have a 100 per cent chance of cancer if they eat, er, 20 per cent more sausages?' And some, who love sausages, say that it's all lies and statistics, but perhaps a few say, 'Oh my God, it's a sausage, stay away!' and they hardly believe it, but maybe they should believe it, and some people tell them they're just stupid, but they don't feel stupid they feel fed up, and still they haven't much idea what it really means, and in the end they say 'Oh, sod it, let's have another sausage.'

---

*Exactly the median birth weight recorded for children born in England, according to the latest available data, equal to about 7lb and 8oz.[2]
† For an explanation of what the 20 per cent increase in a risk does mean and how it's calculated, see the discussion in Chapter 4, 'Nothing'.

Forget all this. We can do better. Norm faces a life full of such fears, often conveyed with about as much clarity. Can we ever calculate his precise fate? No. Obviously. No one knows the future. But as Norm grows up he can learn about the recent past – as with the body count for heart disease – and then extrapolate the average risk into the future and use this as a guide for his own life. This sounds imperfect but reasonable. In practice, risk in the telling is often a mess. Yet so far as the basic body count goes and what it means to our everyday lives, it could be easier. That, at least, is one hope for this book, and for Norm too. Although danger isn't only something people fear and avoid, and maybe even sausages taste better if you think they're on the wild side. Either way, whether you seek danger or avoid it, we will try to make the numbers simple.

Our main technique will be a cunning little device called – by someone[3] with a wicked sense of humour – a MicroMort, which is a 1-in-a-million chance of death. MicroMorts are cheery little units that help us see danger in terms of daily life. They are risks reduced to a micro or daily rate on a consistent scale. The idea starts with exactly that: an ordinary day in the life of any average person, like Norm.

How risky is it for Norm, or for you maybe, to get up, go about your daily routine, do nothing particularly dangerous – no wing-suit flying or front-line duty in Afghanistan, just the ordinary – then come home and go to sleep? Not very, as you would expect.

True, you might lose your life under a bus rather than just your ice cream, slip fatally in the bath or be murdered in a mistaken gangland revenge attack with a power tool, but it's not likely. It is a pretty micro risk. We know this. In fact, we can count the bodies and put a number on it. Typically, about 50 people in England and Wales die accidentally or violently each day by what are known as external causes.[4] * Since there

---

* More precise figures: according to the Office for National Statistics, 18,000 people died from 'external causes' in England and Wales in 2010. That is all those people – out of the total population of 54 million – who died from accidents, murders, suicides and so on. This corresponds to an average of 18,000/54 = 333 MicroMorts per year for each person, or about 1 a day. It is not a perfect benchmark, not least because of the question of how to treat suicide, which is not quite an external event but is categorised as one. But it gives a reasonable

are roughly 50 million people in England and Wales, this means that about one in a million gets it this way, every day. Not a lot, as we said. And even though you don't know for sure if you will be one of today's 50 or so who die from external causes, you're probably not lying awake worrying about it too much.

So this daily risk is about 1 MicroMort, a one-in-a-million chance of something horribly and fatally dramatic happening to Mr or Ms Average on an average day spent doing their average, everyday stuff. One MicroMort, in other words, is a benchmark for living normally. You have experienced this, often. What's more you survived. Congratulations. There you go, one MicroMort, today, tomorrow, every time.

Of course, this is just an average, and who except Norm is average? Some people are too timid to step out of the front door, while their neighbour revs the motor bike on the way to a base-jump. What *Your* risk is (that's You reading now), is a much trickier question that we will come to later. For the moment, please imagine Yourself to be average.

We will also have to assume that recent data give a reasonable idea of the future. In other words, we will move smoothly between historical *rates* – how many people died out of each 1 million? – and future average *risks* – what is the average number of MicroMorts per person? But by 'future' we only mean the next few years – who knows what will happen after that?

With MicroMorts at the ready, the rest becomes relatively easy. How will you spend your daily MicroMort (MM)? If you ride a bike for 25 miles, that's your daily ration. Or you can achieve the same by driving 300 miles, also equal to 1 MM (remember this is an average – 1 MM goes further on motorways). Or you can take a few more risks and add to your daily MM dose.

The joy of the MicroMort – if joy is the word – is that it makes all kinds of risks comparable on the same simple scale. And there are plenty of risks about. For instance, have you in the past ever been born? Do you expect to give birth? Do you now or will you ever drive or fly? Take drugs,

---

approximation to the hazards of daily living, and – provided we keep everything on this same scale – it is all reasonably comparable.

including alcohol or painkillers? Ride a horse or bike? Climb Everest? Work down a mine? Climb a ladder? Spend a night in hospital? Have you or your children ever had a jab? Do your toddlers put small plastic toys in their mouths in flagrant violation of the clear written warning on the packet? What's the risk that an asteroid is right on target for you?

All of these, and every other acute risk can be measured in Micro-Morts (MMs). For example, the risk of death from a general anaesthetic in a non-emergency operation in the UK is roughly 1 in 100,000,[5] meaning that in every 100,000 operations someone dies from the anaesthetic alone. This risk is not as intuitively easy to grasp or compare as it could be. But we can convert it into 10 MM, or 10 times the ordinary average risk of getting through the day without a violent or accidental death, or around 70 miles on a motor bike. We can show you, for example, that every two shifts working down a mine not so long ago in the UK was, on average, about the same risk as going sky-diving once today: about 10 MM. A day skiing? An extra 1 MM, the same as an extra average day of nothing much. Anna would be reassured.

But at the extremes, one MicroMort is also the average risk incurred every half-hour serving in the armed forces in Afghanistan in a bad period, 48 times more dangerous than average everyday living. Or it is the risk incurred by the aircrew of a Second World War RAF bombing mission over Germany in around one second.*

A MicroMort can also be compared to a form of imaginary Russian roulette in which 20 coins are thrown in the air: if they all come down heads, the subject is executed.† That is about the same odds as the 1-in-a-million chance that we describe as the average everyday dose of acute fatal risk.

---

* Between May and October 2009, out of 9,000 UK service personnel in Afghanistan, 60 were killed.[6] This works out at around an average of 47 MicroMorts a day, or 1 MicroMort per half-hour. Between 1939 and 1945, 55,000 bomber crew were killed in 364,000 missions. With an average crew size of around 6, that is around 25,000 MicroMorts per mission, or roughly 1 MicroMort per second.

† The chance of this happening is a half (heads or tails) times a half, repeated 20 times (1 in 2^{20}), which is roughly equal to 1 in a million.

Figure 1: **Some MicroMorts**

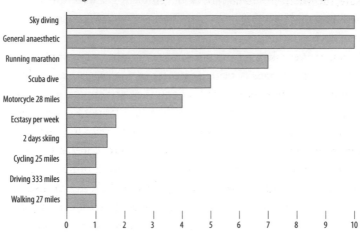

Average MicroMorts (1-in-a-million chance of death)

See Figure 36 in Chapter 27 (p. 292) for lots more and details of how they're done and where they're from.

While we're on the subject, take a moment to decide whether you would be willing to play this game of being executed if 20 coins all come up heads and if we paid you, say, £2 a go.

You wouldn't? £2 is not enough, you say. So how much money would you want in return for accepting a 1-in-a-million risk to your life? In other words, how much is your life, or a 1-in-a-million threat to it, worth?

We can form an idea of how much governments are prepared to pay to save you from a MicroMort by looking at the Value Of a Statistical Life (known in the trade as a VOSL). This is a real-world concept used by governments to determine, for example, which road improvements to make. If a new junction is expected to save one life, then in the UK we will pay up to £1.6 million for it.[7] Therefore the government prices a MicroMort at one millionth of this, or £1.60. Were you willing to play the Russian roulette game for £2? No? The government thinks you're over-stating your value.

We will present risks in a variety of ways, but we will use MicroMorts often. The MicroMort describes acute risks, those that hit you over the head with a 'Thank you and goodnight'. Later, we will introduce another measure, the MicroLife, to talk about longer-term hazards: chronic risks of the kind that slip slowly into the bloodstream and build up over a lifetime, such as cigarettes, diet and drink.

Both measures take a few liberties and make some sacrifice of precision in exchange for ease of use as a daily measure of the hazards of life: one number, easy to compare with every other, rather than a muddle of percentages. Sometimes we will show the calculations involved, but we will tend to put these in the notes for those who want to skip them.

It is partly on the sense of proportion from MicroMorts and Micro-Lives that Norm will eventually stake his hopes for a fulfilled life, in which knowledge and comparison will guide him, help him realise his full potential to become the flourishing, uninhibited Norm he was always meant to be. So far, so reasonable – if reason is what counts. In time, we'll see. First, there's someone else we'd like you to meet.

2

INFANCY

F ROM THE DAY PRUDENCE WAS BORN, her mother was never
not afraid.* She became protector, haven, she-wolf, sniffing danger
like a forest animal. Other children were germy, snotty, clumsy, other.
Fungus and barbed wire, on legs.

Watch her now in the loo, with thighs of iron, paisley skirt hoisted
to her armpits, hovering. See how vigorously she then washes her hands.
Meanwhile, notice Prudence in the pushchair reach out with infant
curiosity to touch ...

'No, no, no! Dirty Prudence. Don't ... ever ... Over here ...'

'Mama.'

'Hands ...'

'Huh ...'

'Oh you haven't? You have. All right. Legs up. Nappy ... don't touch.
Whoops. Where are the wipes?'

Wipes. Trusted companions. First line of defence in the war with

---

* 'Once she was born I was never not afraid. I was afraid of swimming pools, high-
tension wires, lye under the sink, aspirin in the medicine cabinet, The Broken Man
himself. I was afraid of rattlesnakes, riptides, landslides, strangers who appeared
at the door, unexplained fevers, elevators without operators and empty hotel
corridors. The source of the fear was obvious: it was the harm that could come to
her.' Joan Didion, *Blue Nights*[1]

dirt. Prudence would learn from an early age never, *never* to take peanut butter sandwiches onto school premises, *never* to mix the knives used for raw chicken. Just as her mother worried for the unaware, pitied them their lack of foresight and the risks they ran. Invited to join a book club, she nailed instantly the danger others missed – that if she lay awake to read at night, her husband might think she had the energy for sex.

The headline in her morning paper, '20 physical signs that you are seriously ill', meant Prudence's Weetabix must wait while she read it, twice, and self-diagnosed. Was that a new mole on Prudence's leg? No, a spot of mud. Where were the wipes?

Follow mother and daughter now as they emerge from the loo into the coffee shop, where an old friend sneezes. See how the mother leans back from the table, lips zipped, breathing as if air hurts, trying to disguise the sweep of her hand over Pru's mouth and nose. She's thinking of disinfectant hand-wipes.

'You need perspective,' a male friend once told her, and offered to show her some probabilities.

'Thank you,' she said, 'but numbers aren't the point.'

'What?'

Bad things happened in the world, unspeakable things. The thought of 'what if?' – what if the worst happened – beat the numbers game every time. 'What if' turned any risk, however small, into Prudence ... screaming.

'What if I show you it's one in a million?' he said.

'No good.'

'What?'

'The problem's the one.'

'One in a million is good.'

'Not to the one. Not to the one.'

Especially if the one was Prudence. Was her childhood home child-friendly? It was. Including the garden ponds or water features? They were dug up. Did her mother know how to react in an emergency? She learned. Had she chosen safe equipment and furniture? She had. There was safe and healthy eating, safety on holiday, sleep safety, bath safety, safety on stairs and sofas, little objects and little bones (easily broken), scalds and burns, suffocation and drowning. Parenting magazines kept her alert. 'Neglect' was almost the worst word she knew.

She also knew that some terrible twist of nature, an undiscovered illness, say, might still take Prudence. But short of that, and with proper care, she would grow up healthy, to outlive the reckless and misguided who drove too fast, ate badly, ignorant of cholesterol and acrylamide, grew fat and, careless of gender-bending pollutants out there, left electromagnetic devices switched on in bedrooms and even had their children vaccinated. Or was it that they didn't have their children vaccinated? Bugger, what was the latest? And Prudence was due a jab. Sometimes the threat seemed to come from both sides at once, like two-way traffic.

Children on a winter's day meant a slip hazard; summer meant wasps and sunburn, with an eye for jellyfish on the beach and tsunamis on the horizon, when even restful moments with the *Mail* and a mint tea required a watchful eye over Pru's play, which certainly ought not to stray beyond the line of the deckchairs. An advertisement on the train, 'Don't risk bad breath!' was on an all-embracing continuum with terrorists: 'In these times of heightened security, please ensure you keep all your belongings with you. If you see a suspect bag or package ...'

This was the world Prudence would inherit, a world of hazards, threats, risks or symptoms, in which numbers hardly mattered but a scary story might, in which fear was the price she paid for love, in which Prudence, hurt, was an infinity of pain and guilt for her mother too – as she felt again on the way home after accidentally banging her daughter's head on the car door.

PRUDENCE'S MOTHER IS RIGHT to this extent: being a baby is high-risk – relatively. The hazard in the first tender year of life is roughly that of riding 30,000 miles on a motor bike, once around the world.* Imagine your baby on a Harley. Feel safe?

Those who survive won't face the same level of annual hazard again until their mid-50s. Under-1 is a dangerous age, comparatively.

---

* Assuming it's all on British roads. The risks on an Afghan pass are probably higher.

But you probably knew that. Children's vulnerability is a cliché. First helpless, then a dicey mix of naive and curious, they're asking for it, aren't they? It's easy to see them all on the same high wire until they grow up.

Easy, but misleading. This splurge of fear is too crude. It masks a stark difference in risks that change radically in only a few years. Prudence's first year is relatively risky all right, but most of the risk is squashed into the first few weeks, and if she makes it to her first birthday, as the vast majority do, her annual risk plummets – from 4,300 MMs in the first year to fewer than 100 MMs a year for a seven-year-old, or about a quarter of a MicroMort a day from all causes. Believe it or not, this makes seven the safest age of all to be alive – far safer than life for mum and dad.[2]

So infancy and childhood go in short order from one extreme of life's acute risk to another. Babies don't do much, but they do live dangerously, briefly, compared with others. As soon as they become young children, they get up to all sorts, safer from death than anyone.

This is only fatal risk, and there's plenty of damage you can experience short of the big one. Even so, the change in fortune is stark. Anxious parents have something to worry about in year 1, most of it in the first weeks, and then ...

But does their anxiety fall with the risk? Or do they worry away anyway, like Prudence's mother? If so, maybe it's watchfulness that keeps children safe, in which case worry works. Certainly, she thinks so. A danger foreseen is half avoided, an old proverb says. Or maybe once parents start to worry, they can't stop, and it takes about a fortnight for paranoia to set in.

But there's another reason her mother is frightened: not just love but also fear of blame. Danger can feel worse if you think someone is at fault when things go wrong, if someone else 'did it'. Then there's an innocent victim, and an accident is more than bad luck, it can become a grievance. If you are the one doing the looking after, when protection is your duty as well as your care, if you are not just loving but also the one who'll be blamed, then danger is darker – and guiltier. 'And where was the mother ...?' someone will ask.

So if something happened to Prudence, how would her mother feel? Amidst everything else, rightly or wrongly, she'd feel she hadn't been there. You could tell her it was bad luck all you liked. She wouldn't hear. Probabilities don't mention this – the human pain.

But don't judge them too harshly. This indifference can be a virtue. Probability is uncaring, it's true, but by the same token it is often un-blaming. Not all the feeling in stories is benign – and so probabilities can be kinder than stories, more forgiving, less eager to find human agency (a particular person who 'did it') and happier to admit their uncertainty about exactly how the accident happened.

That's partly what probability is – a statement of uncertainty about cause-and-effect or naming the guilty. So, being an emotionally chal-lenged number can have its humane side too, if only by omission.

What the numbers primarily say is that the most serious risk of death in infancy is brief, as we say, and also that it has plunged. Infant mortal-ity is a good indicator of the change in social conditions, and strongly influenced by transient famine and epidemic. It tells its own story of breath-taking progress.

Throughout ancient history perhaps 30 to 40 per cent of babies would die before their first birthday, what we might call the natural rate of infant mortality. By about 1600 in England, the figure had roughly halved, and remained at about 15 per cent from the mid-1800s.[3] If this were the rate today, it would mean well above 100,000 infant deaths a year in the UK.

Fortunately, it improved again, dramatically. By 1921 the rate had almost halved for the second time, but this time the leap took just one generation. Soon after the Second World War, it had halved again. By 1983 even that rate had halved again and then halved again, and then by 2012 it more than halved again to stand now at about 4 per thousand.* The change has been nothing short of astonishing, death slashed back, again and again. In much of the world the biggest historical cause of

---

* Although infant mortality had fallen sharply up to 1921, it still accounted for a staggering 82 babies out of every 1,000 born in 1921. Then it dropped sharply in only a generation, to 46 in 1945. By 1983 it was down to 10, and it is about 4 now.

early death has been so reduced that what was once ordinary is now exceptional. Has there ever been a more radical reduction of risk in human history than the risk of infant mortality in developed countries?

How do those numbers make you feel? Privileged by progress and reassured for your children? Or still anxious?

For though the risks have reduced massively, they are still around. By 2010 in England and Wales there were 723,165 live births, more than one a minute, or close to 2,000 a day: enough to fill a very large and very noisy cinema.* Think of Prudence as one of them. So far, so good. But she had to survive several scary moments along the way.

The first came before the first breath. If you look around that cinema, you notice ten empty seats. These represent the stubborn toll of still births, around 1 for every 200 live births, a rate unchanged since the early 1980s. While other countries have continued to make progress, the UK has not. This is a puzzle, and it is disturbing.

Next comes survival in life. Of the 2,000 live babies born each day, around 5 die in the first week, another 1 before a month is out, and 3 more before they reach their first birthday.[4] The total risk during that first year from a variety of causes is, as we say, 4,300 MicroMorts. That is how we arrive at the 30,000–miles-on-a-motorbike equivalent. If 1 MM equals riding a motorbike for about 7 miles, then 4,300 MMs x 7 miles = 30,100 miles.

We'll take the dangers to Prudence that make up this total one at a time. First, the largest single category: congenital disease or prematurity. Very tiny babies struggle. Of the 4,000 babies born in 2010 who weighed less than 1kg (a bag of sugar), 1,200, or 30 per cent, did not make it to their first birthday.

As Prudence was not premature or congenitally ill, the 30,000-mile motorbike equivalent risk falls sharply to about 16,000 miles, from 4,300 MM to about 2,300 MM.

But then comes the next danger: 202 babies, four a week, who died after the birth went wrong. It is a continuing controversy whether it's safe to have your baby at home or in hospital (at least, since hospitals

---

* 2,000 is the capacity of the Empire cinema in London's Leicester Square.

Figure 2: **Causes of death in infancy, England and Wales, 2010**[5]

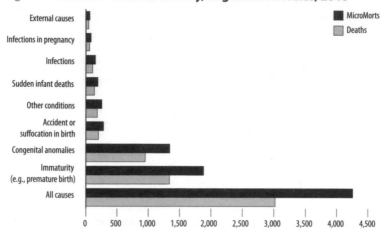

learned the lessons of hygiene; see Chapter 11, on giving birth). People who give birth at home are generally better off financially, and that tends to be associated with higher rates of survival. They also tend not to be having their first baby, which should also be safer. Even so, the infant mortality rate is the same for home births and hospital births in England and Wales.

A recent study of 65,000 'low-risk' births showed that women giving birth in a unit run by midwives were just as safe as in hospital but had far fewer caesareans and more 'normal births'. If it is not your first child, then giving birth at home is as safe as in hospital, but for first-time mothers there was around double the risk of a serious problem, and nearly half had to be transferred to hospital.[6]

The final big danger to Prudence – and everyone else – is the 136 Sudden Infant Deaths (192 MicroMorts) added to those cases that are 'unascertained', to leave a distressing group that is officially called 'Unexplained Deaths in Infancy'. Since the launch of the Reduce the Risk campaign in 1991, which overturned the received wisdom of leaving babies to sleep on their front, the number has dropped by 70 per cent. But there were still 279 in 2009 in England and Wales (400 MicroMorts in the first year), around 1 of the 2,000 babies born each day.[7]

These mysterious deaths are around 50 per cent more common in boys, more common in winter, and the risk is five times higher for children of mothers aged under 20 (1,230 MicroMorts in the first year) than over 30 (250 MicroMorts). So all in all, as a girl, not premature and with a mother in her 30s, Prudence was among the safest at this – relatively – dangerous age. Not that her mother would have been reassured.

That is in a developed country. What if she had been born elsewhere? International comparisons for infant mortality are tricky. Some countries don't include in their infant mortality data the tiny premature babies that we have seen are at such high risk. Lack of good registration also means that the rates have to be estimated from surveys that ask households if they have had a child under five who has died, then use statistical models to estimate infant mortality.

The admirable UN Inter-Agency Group for Child Mortality Estimation[8] puts all these data together and estimates that, over the whole world, the average risk faced by a baby in the first year is around 40 per 1,000 (a chilling 40,000 MicroMorts), about the level in England and Wales in 1947.

But, like so many averages, this obscures massive variation. Bottom of the league table are Sierra Leone and the Democratic Republic of Congo, with rates of around 119 and 112 per 1,000 respectively, the English rate in about 1919. Ethiopia comes in at 52 (the English rate in 1938), India 47 (1945), Vietnam 17 (1973), with the USA at 6 (1997). Cuba's 5 per 1,000 is a fraction behind England, while Finland and Singapore are down to 2 per 1,000,[9] about half the rate of the UK. This suggests that we cannot separate the world into 'us' and 'them' – 'they' are mostly like us a generation or so ago, and catching up fast.*

The line in Figure 3 shows the dramatic improvement in the UK's infant mortality rate since 1921, measured in deaths per 1,000 live births before one year old. We've also shown the current position of a selection of other countries. Thus Cameroon in 2010 was about where the UK was in the early 1920s.

---

* See Hans Rosling's Gapminder project (www.gapminder.org) for brilliant demonstrations of this.

Figure 3: **Infant mortality rate per 1,000 live births, UK historical trend, with current position of selected countries**[10]

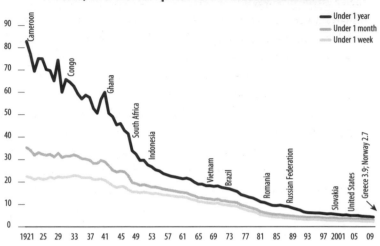

The Millennium Development Goals were set up by the United Nations in 2000, and the fourth goal was to reduce infant mortality by two-thirds between 1990 and 2015, from 61 per 1,000 live births to 20. The current level of 40 shows great progress over twenty years, but means the overall goal is unlikely to be met, although some countries have made giant strides: for example, Malawi has gone from 131 to 58 per 1,000, and Madagascar from 97 to 43, both of which are 56 per cent reductions in twenty years.[11] If you have the impression that the developing world makes little progress in human welfare, think again. The number of babies who died before their first birthday is estimated to have fallen from 8,400,000 in 1990 to 5,400,000 in 2010 – an astonishing improvement but still 15,000 a day, 600 an hour, 10 a minute, 1 every six seconds.

Here's an odd question: is the death of a baby just nature's way, as Prudence's mother fears, or is it unnatural? The distinction matters. Many people feel that unnatural risks – risks associated with modern living, such as travel, technology or obesity – are worse. Humans created them because enough of us liked the benefits from cars, nuclear power or cakes. But if what we create also messes with nature, well, some say, what do we expect except our comeuppance? Better to go by a lightning strike than a downed power line, as one piece of research put it.[12] Maybe these unnatural risks feel worse because this is what happens when humans get above themselves, or maybe they feel worse because someone else imposed them on us. In contrast, we might call the death of a very small child more like supreme bad luck, an act of God or nature, primitive or given, especially when illness is the cause.

'Natural' risks are often more tolerated. But it's an awkward attitude, and this chapter tests it to the limit. For what greater fear is there than that of seeing your child die? The narrative is so disturbing that even TV uses it sparingly, while portraying adult death on an industrial scale.

Disease is natural and slaughters people in the millions. Should we tolerate it because it's always been around? And yet 'unnatural' remains a potent criticism, even though humans brought a dramatic decline to the natural rate of infant mortality, sometimes with nothing more unnatural than germ theory and soap. Few people think we should not have messed with it.

Although the anti-nature argument can be taken too far: does it mean technology is always good? Obviously not, even for children's health. Neither side wins every time. The natural/unnatural risk argument rumbles on around home birth (and vaccination – see Chapter 6).

Is this odd preference for natural risks irrational, when natural risks can be the most devastating of all? Strangely, maybe not. 'Unnatural' can be a way of saying that you don't like a risk for other reasons. Maybe what you mean is that an unnatural risk is when people or big business get up to no good. Maybe you think someone is making too much money or seeking too much power. 'Natural' and 'unnatural' might be bound up with tricky ethical or moral feelings about the way a whole society behaves, nothing to do with hard and fast distinctions in nature itself. This doesn't make these feelings irrational, but it does make them complicated. Risk is often like that: ostensibly about danger, but really packing a whole lot of attitude about a whole lot more.

From thinking of one child, our own, and trying to imagine the loss, to thinking about the same loss multiplied more than 5 million times every year, then of another 3 million who now survive thanks to economic development, technological progress and the medical control of disease, seems to leave little room for sentiment about risk and nature. But the sentiment persists. What does that say about us?

That risk, even a risk like this, is a small part of a sprawling human calculation about what's right and how to make progress, a calculation that's sometimes political, sometimes moral and sometimes, possibly, simple vanity.

3

VIOLENCE

HARRY THE HAWK's gimlet eyes policed the city's criss-cross streets on behalf of pest control division – 'Rat-Swat' – of the municipal department of environmental health. High above, he twitched his feathered wings and rode the air, watching.

People, about their business. Vehicles, moving and stopping. Trees in the breeze. Children in danger.

There by the park, for instance, walked nerdy Phil, now in long trousers, not far from the older lads hanging out in the park. Phil stopped to stare at a puddle. The puddle fizzed. From beneath the broken words 'Central Electricity Board' on a half-sunk and vandalised iron cover came pops and flashes.

'Idiots,' said Phil, a tad too loud as he stepped around the water. 'They've actually electrified the pavement!' Harry saw the lads walk over. There was shouting, a scuffle, a knife. Phil fell – with a splash.

Across the street, Harry could see Mikey leaning on the railings. Mikey had been there a good 15 minutes, had made the call on his Blackberry to report the puddle shortly before a stranger's hand scooped up the small rucksack with his laptop, GCSE textbooks and notes that he left leaning against the post-box.

'Why is he wriggling in a puddle?' said Norm, who was nearly three and a bit, as he stopped beside Mikey, who had turned away from the railings and was now crouched and hiding behind the post-box tapping the phone again and not answering him and saying a word Norm thought might be not a very nice word.

'Are you hiding?' Norm said.

He was a bit frightened before, when daddy went and didn't come back somewhere and told him to stay here but he couldn't remember where here was. Now it was cold. And the man with the phone looked at him in a funny way. Norm wondered if he was a stranger.

'Because we don't want him getting a chill,' said Mrs Assabian, just along from the puddle at number 38, as she wrapped Artemis in his Dolce & Gabbana fleece jacket and tightened the buckles under his little tummy. 'Now go,' she said to her nine-year-old daughter Jemima, whose other bruises had largely healed, 'or you'll get another' and clipped on the lead while shoving them down the steps, turning back inside and closing the door.

Harry registered in fine detail the movement of these two small figures, one upright, one horizontal. Horizontal had four legs. This was important. It scuttled from left to right, intermittently jerking forward, a pattern Harry recognised by instinct, honed by reward. He watched. Then he swooped.

Bloody big rodent, thought Harry, as he sank his talons into the red coat with the strange markings and began to haul the vermin aloft. But the vermin was stuck. Harry flapped harder.

Mrs Assabian's daughter, by now in the High Street at the window of a shoe shop, felt her right arm wrenched to the left behind her then yanked into the air. Twisting, she met Artemis at eye level, his paws dangling, snared in the claws of a beast beating the air.

'Noooo!' she screamed, 'Get off, get off!'

She pulled and screeched, fighting as she had learned to fight other blows. Harry screeched too, louder, and flapped harder. A male passer-by spotting a once-in-a-lifetime chance of tug-of-war for an airborne chihuahua, joined in, laid a hand ineffectually on the leash and absolutely screeched.

Shoppers turned – and screeched. Artemis screeched, as best 3 lb. of dog can, and then, with a rip, fell to earth. Harry flapped away, towards the river.

Where the crescendo of screeches carries the short distance over the city to find the four-year-old Prudence in the car seat as her mother drives them home from a trip to the big city. The traffic noise surges around them. Mummy hoots the hooter. From somewhere behind them, another hooter blares, livid and too long.

'What's that, mummy?'

'Silly man.'

A small van behind them whines up through its gears. The van is veering and tilting insanely across the lanes. It squeals and stops in front of them. They brake hard, inches shy of a crash, and the horn blares again.

Blind to the traffic that checks and veers around his van, Van flings open the door and jumps out, engine running. She tries to pull away, but they lurch and stall. Head down, she twists at the key, pulls the wheel hard left, lurches, stalls again, twists, starts, revs high in panic.

Too late. Van has swept around his bonnet and as they begin to drive, leaps at the side of the car, clinging to the rim of the roof with one hand, feet unseen wherever they can hold.

Tongue out and leering, but silent, Van's face is pressed to the driver's-side window, his other hand a fist, beating at glass, door, windscreen. She accelerates. Van is still there, a limpet, an alien, a beating fist. He has to jump. He must. He holds on. Still he holds on, then jumps backwards – and spins onto the road.

It has taken just seconds. Not a word, not a shout, plumes of breath in the cold, Van saunters to his van, settles in as he moves off, reaches for the still-open door as it waves out against the turn of the van, pulls it shut and drives off.

Prudence is silent. So is her mother, staring straight ahead as she grips the wheel. Never complain, never look, never speak out, keep away, strangers, mayhem, accident, danger, madness, hatred, death.

High above, Harry circles.

They drive home. In the column asking if the parent gives permission for the school day trip, Prudence's mother circles 'I do not'.

FOR A PRICE, Norm's parents could have his genomic DNA isolated and quantified using agarase gel electrophoresis, in which 'a region of gDNA is selectively amplified by polymerase chain reaction (PCR)' for a home DNA storage kit.

Why? In case he gets lost – as he does. You could do the same for your own children. At least, that's the most innocent motive. The darker thought, the thought of what DNA is used for, is to detect crime. And that means abduction, or worse. The sooner the police have these DNA details the better, say the instructions on the tin. You can also have your child's tooth-print recorded on an arch-shaped thermoplastic wafer. Let's not dwell on the moment that comes in handy. Fingerprints, identity bracelets and recent photos are often recommended.

Abduction is a scary thought. Does this kit make it any better? Would you feel happier for Norm knowing that we had his DNA in a tin? That's the problem with fear, a cruel, insidious thing: finding ways to alleviate it means thinking about it.

Could we ignore it instead? The chance that a stranger will try to abduct Norm or any other young child is on average extremely small; the chance that the stranger will succeed is 80 per cent lower than that. The chance that someone will murder your child is smaller again; for murder by a stranger, even that risk is cut by about another three-quarters. If you are a parent, by far the biggest violent risk to your child, on average, is you.

Even including the risk from parents, the risk of death from all causes for young children like Norm is the lowest it has ever been, and this is also, as we saw in the last chapter, the age of lowest risk in an average human lifetime. That is, based on the most recent data, early childhood is the safest age to be alive, at probably its safest moment in history.

So what? Bad things happen, as Pru's mother says – and she is right. Our fiction is an absurd batch of shocking dangers and calamities – deliberately so – especially in the space of a few minutes and a half-mile radius. But they are all loosely based on real events. The electrified pavement and the curious incident of the hawk and dog did happen, within a few hundred metres of one another. Every so often a child dies in an animal attack, though it tends to be dogs not birds of prey. The road rage happened too, though all the characters have been changed here and some events altered. Some may remember, more sombrely, the story of

Ryan Jones in 2007, an 11-year-old shot dead in a pub car park in Liver-pool, an innocent in the way of a gang feud. Or Baby P, as he was known, a 17-month-old boy who died after suffering more than 50 injuries over eight months at the hands of his mother, her boyfriend and his brother.

These were horrifying cases. Our problem is that horrifying cases dis-tort our sense of the odds. Precisely because they are rare, these events stand out. The word 'salient' describes risks that catch the eye. In this context, salience means something like the way a light seems dim or bright depending on its background. So one appalling example now and then against a background of low odds stands out more than if the danger is common. It is a troubling trade-off: just as we pay for trying to avoid the worst with being reminded of it, so we pay for rarity with sensitivity.

So is it despite or because of the fact that the chance of violence to young children is so remote, that parents have grown – according to some writers – more anxious about their children's safety?* The salience paradox is one putative explanation for this split between low odds and high anxiety. The safer life becomes, the more terrible seem the extreme exceptions.†

A question for everyone is whether this gap between the odds on disaster and how anxious we are proves we're paranoid and irrational, or that the data miss the point, or both. To the rationalist who says: 'But don't you realise how rare these things are?' there is, for some, a paradoxical answer: 'Yes, which is why they are so alarming.'

Here are some of the numbers. The risk of homicide is lower for children aged under 15 than for any other age group, at between about

---

* A growing fear for children has been dated to the 1960s, with the emergence of battered child syndrome, followed by runaways and Halloween sadism in the late 1960s, sexual abuse, child pornography and child-snatching in the 1970s and missing children and ritual satanic abuse in the 1980s.[1]

† Although there is a problem with this 'rare is worse' feeling. If it also suggests the obverse – that 'more common is not so bad' – then it is not so far from the attitude that life is cheap in developing countries: 'So many people die, can it hurt much to lose another?' Readers must decide whether a fatal event feels worse if it is also more unlikely.

Figure 4: **The annual chance of being murdered, by age, in MicroMorts (averaged over the three years, 2008–11), England and Wales**

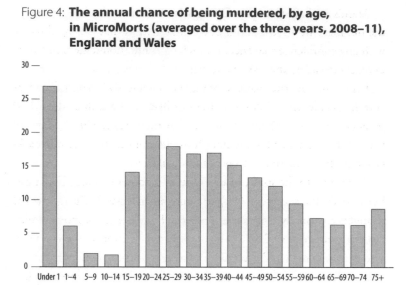

2 and 5 MicroMorts every year. So it takes a year before a child is exposed to a murder risk equivalent even to the normal hazard of acute death from non-natural causes that an average adult is exposed to every few days.[2]

That's children. For infants, it's a somewhat different story. Those aged under one are more likely to be murdered than any other age group. But the risk is still low compared with the daily hazard of average living. At about 26 MicroMorts a year, or about 0.07 MM a day, it is roughly one-fifteenth of the adult baseline everyday risk of death from all non-natural causes of 1 MM a day.

But still, as Pru's mother says, it happens: about 46 child murders a year (on average), out of around 10 million children under 15. A more homely, more human and more sinister way of describing how often children are murdered in the UK is about once a week.

But if you can rule out murdering your own children, you cut out about three-quarters of the average risk for all under-15s, which all but wipes it out for most children in this graphic, and reduces the murder risk for under-1s from 26 MMs to about 5, lower than for almost all adults. And yet still it will not be zero.

Trends are harder to spot. Disregard what you read to the contrary,* the numbers are so low and prone to fluctuation that it is hard to say with any confidence that there is much of a child homicide trend at all in recent decades, one way or another.[4]

For abduction, the chances are higher. But a study in 2004 found, as with homicide, that children were most likely to be abducted by their parents (141 cases, about 90 per cent of them successful).[5] Many of these were by partners, sometimes from different racial or ethnic backgrounds, disputing the custody of their children.

According to the study, strangers attempted many more abductions but were far less successful (364 attempts, 67 successes). The motives of strangers are usually assumed to be sexual. For these cases, short of murder, the same study said that 'in all offences where information was available the abducted child was recovered within 24 hours of being taken'.

Homicide and abduction count only the worst of what happened. But the difference between what does happen and what might is one of the ways in which risk calculations seem to some people to miss the point.

The figures count actual abductions or obvious attempts, not malicious intent or vulnerability. So let's try to capture the sense of potential risk rather than only the real incidents, and count all the cases of missing children – as just one among many possible ways of measuring vulnerability.

Now the numbers explode. In 2009/10, according to the Child Exploitation and Online Protection Centre, there were an estimated 230,000 'missing person incidents' for children under 18.[6] But again, most return within 24 or 48 hours, more than two-thirds seem to go missing because they decide to, and the incident count can include the same person who goes missing more than once.

---

* In 2010 it was widely reported that the risk for all children under 15 had fallen sharply over the years, to the extent that the BBC and others repeated claims that it was about 40 per cent lower than in the early 1970s.[3] This is unlikely. Those figures are largely a result of changes in the late 1970s to the way cases were recorded while still awaiting the outcome of criminal proceedings. This problem does not affect the MicroMort figures we use for homicides, which are based on consistent methodology, and although the odds we have stated will be wrong in the sense that they are not precise, they will be right enough.

Even so, as the report says:

> When a child goes missing, there is something wrong, often quite seriously, in that child's life ... Abuse, exploitation, and risk to life are the most concerning of all dangers that children face. Other risks include violence, criminality, and loss of potential due to lack of school attendance or other education, lack of economic wellbeing, sleeping rough, hunger, thirst, fear and loneliness.

But why count only missing children? Some would count all children everywhere for the measure of who *might* fall victim.

Part of our anxiety is who might be out there with a vile motive. Until these people commit an offence, that is hard to know. Afterwards, the authorities try to track all offenders who are not in prison. Since some are tracked for many years, even for life, the total number like this has risen relentlessly since the tracking came in a few years ago. That does not imply a growing threat; it simply reflects the fact that we have only just started counting and the numbers are adding up. Not all these tracked offenders have offended against children, but those who have and are still considered a risk are included with others in the table reproduced below, under what are known as multi-agency public protection arrangements (MAPPA).*

In 2010 there were more than 50,000 offenders tracked and managed by MAPPA, which sounds truly alarming. The number at the most serious level of sexual risk was 93. (These are known as Level-3 offenders, the toughest to deal with and control, including those who hit the news.)

And supervision doesn't always mean prevention. More than 1,000

---

* MAPPA gives all these figures in more detail.[7]

Management levels	Registered sexual offenders	Violent offenders	Other dangerous offenders	**Total**
Level 1	35,665	12,985	–	48,650
Level 2	1,467	744	438	2,649
Level 3	93	56	41	190
Total	37,225	13,785	479	51,489

offenders of all types were returned to custody for a breach of licence or of a sexual offences prevention order, and 134 MAPPA-eligible offenders were charged with a Serious Further Offence. Again, for many people the simple presence, out there, somewhere, of anyone who has a history of harming children is enough, and the vagueness of the threat – you don't know what they might do or who they are – only makes it worse.

Some parents are not that worried. Some worry only in crowds. Some can't bear to let their children play outside. Some worry only when a crime hits the news.

And some say it's all a 'moral panic'. The phrase was popularised in the 1970s in an analysis of the Mods and Rockers of the '60s written by a sociologist, Stanley Cohen.[8] In brief, the argument is that the media overreact to behaviour that challenges social norms, and this overreaction comes to define the problem, even creates a model for others to copy. It is a powerful analysis. But the word 'panic' suggests irrationality, as if you are a parent, and you see another parent lose a child to a violent stranger, and your heart screams, but you must not pull your own children closer. Is that reasonable? Or is it the case that nothing can tell us when proportionate reaction ends and overreaction starts? There is no tape measure – in MicroMorts or otherwise – for the shock of a child violently killed. Was the shock disproportionate in December 2012, when 20 children were murdered by a gunman at Sandy Hook Elementary School in Newtown, Connecticut? Does that shock change if you know that there are roughly 15,000 homicides in the US each year?

4

NOTHING

THE BIG SAUCEPAN on the hob with water in made funny plop-
ping noises. Steam hissed out. The egg inside rattled. The blue fire
underneath was jumping. Prudence went to see.

Usually, mummy didn't let her – and turned the handle away. She'd be
next to you if you did something naughty and she said it was dangerous.

'Prudence, are you there?'

She'd say, 'No' (before the scalds, the shrieks, the inconsolable pain).

'Prudence? Don't wander off, love.'

Today the little girl wanted to pick up the saucepan like mummy
picked up saucepans, and she went to the gas and the bubbles, nearer,
and she reached for the handle, and mummy came in, picked her up and
turned it off.

'There you are! And it's ready.'

Meanwhile, six-year-old Norm was sulking.

'I can't find it!'

His father turned around in the driver's seat. The lights changed.
The car behind hooted. He jolted back in his seat and glowered in the
mirror, thought about sticking a raised finger through the window, as
seemed to be the fashion nowadays, then thought better of it and pulled
away. But Norm was convinced they'd left it behind.

'It's not here!'

His father turned again to reach back and rummage in the pile on the seat, then change a gear, turn again, glimpse and accelerate.

You couldn't blame the driver of the Waitrose lorry coming in the other direction. He had no chance. He scarcely touched the brakes as Norm and his dad veered over. In one awful moment the cab of the HGV loomed and filled the windscreen and Norm's dad popped the wheel over to get out of the way and pulled into the bus stop to look for Norm's Sudoku properly.

'Be careful, Prudence, you might fall.'

'But I might not.'

At home, Kelvin lit a match and dropped it into the ashtray on the kitchen worktop. He watched as a wicked flame got the napkin, but then caught a corner of kitchen roll that was near, then the tea towel. The fire was growing. Kelvin stepped back. He didn't mean that. What had he done? Before his parents in the living-room could separate the smell of their own cigarettes from real smoke, fire had taken the curtains and Kelvin was staring at a sheet of flame. His dad raced into the kitchen to find that pretty much everything flammable had gone up – there wasn't much – and the fire had petered out.

By evening, Prudence's mother stood in the doorway, arms folded, eyeing her husband sprawled on the sofa in front of *Newsnight*. Useless, waste of space he was, her best kitchen knife lying ruined on the coffee table where he'd left it after changing a fuse on a plug because he couldn't be bothered to go out to the garage for a screwdriver. Look at him. Useless, always in the way, with not an interesting word to say, and what could she have seen in him? And ten years of resentments poured through her. Thirty more of ghastly married solitude reared up. And she stared at the knife and felt a despair that flared into a rage so sudden that in one swift movement she strode over and took the knife and felt the strength of vengeance in her right arm for all her murdered hopes as she saw the terror in his pleading little eyes and raised her hand and plunged the knife deep into his chest in one of those trivial fantasies of resentment that came and went in a blink before she sighed and smiled, and went over to pat his head.

'Ready for bed, love?' she said.

IN A COMEDY SKETCH by David Mitchell and Robert Webb, a film-maker (Mitchell) is interviewed about his *oeuvre* after a clip from his latest work: *Sometimes Fires Go Out*. In the film, as a couple watch TV, a small fire in the kitchen– yep, you got it – goes out. That's it.

The interviewer (Webb) says the film has been reviewed as 'unrelentingly real', 'a devastatingly faithful rendition of how life is' and 'dull, dull, unbearably dull' – all those comments, oddly, from the same review.

He introduces a clip from another film: *The Man Who Had a Cough and It's Just a Cough and He's Fine*.

Two Edwardian lovers have a series of rendezvous on a station platform. The man, spluttering, looks more pallid and doomed with each encounter. 'It's just a cough,' he says, stoically.

Except that it is. It's just a cough. In the last scene he's dandy. It is one of the finest comic sketches about probability you'll ever see. But then, where's the competition? The idea is lovingly ripped off in our fiction, with some added examples.

Explaining jokes is a bad idea, and the joke here is simple: stories are about what happens; they're not often about what doesn't. If nothing happens in a story, it is not usually a story, it's a joke. And that, funnily enough, is a problem.

Fictional stories choose what to bring to our attention, or their authors do, and whatever they choose, they choose for a reason, often to prime the reader for what comes next. Anton Chekhov said: 'If in Act I you have a pistol hanging on the wall, then it must fire in the last act.' Or imagine an episode of the hospital drama *Casualty*, the family gathered around the breakfast table, when the old man coughs ...

You know what's coming next. A cough in everyday life is not what you might call a statistically significant event. But you're watching *Casualty*. And so you know that this will be a triple heart bypass.

And you are right. Likewise, a pan of boiling water and a toddler can mean only one thing in fiction, and it must involve screaming. In stories, do fires ever just go out?

Risk is much the same. It is almost always framed by the thought of events, not by non-events, framed by things that happen, not things that don't. As soon as we talk about the risk of heart disease, we are thinking of those who will die of heart disease, not those on the other side of the odds who will be fine. The whole discussion of risk is primed by the thought of the bad stuff that happens. As Chekhov should have said, everything you hear about risk begins with a revolver on the wall.

All this may be true, you say, but how else can we talk about risk? What's the alternative? The alternative is to approach the whole subject with our focus on what doesn't happen, the non-events, the coughs that are just coughs and the fires that go out. Then risk would no longer be about the deadly, unhappy endings, it would also be about your chance of being fine if the risk didn't, in the event, burn down the house.

This is an unorthodox take on the subject, admittedly. You'll look in vain for the headline 'No children killed on the way to school today'. Non-events are by definition systematically neglected in news coverage. Imagine the newsroom conversation that begins: 'Anything interesting not happened today?'

But is the concept so daft? As we said earlier, a probability like 30 per cent implies a non-probability of 70 per cent. There are really two numbers, not one: two sides to the odds. This reminds us that the way risk is usually framed places one number in front of the other, with no equivalent word for thinking of them the other way round. What is the opposite of risk's emphasis on events? Not safety, exactly. That doesn't quite describe the nothingness we're after. The dictionary and thesaurus are not much use, offering a meagre list of synonyms for non-event such as 'damp squib', which hints at how alien the concept is. We scarcely have words for it. But that's because it is unfamiliar, not because it's ridiculous. How would a newspaper describe the thing that didn't happen, if it ever did? Would it see the point?

Take a real example. 'A daily fry-up boosts your cancer risk by 20 per cent', said the *Daily Express* in early 2012,[1] about the effect on pancreatic cancer of processed meat. Not only does the *Daily Express* put the cancer revolver on the wall; when it also says that the risk is 'up', it caresses the trigger.

This '20 per cent up' is a calculation based only on the things that happen. It begins with people who have or will have cancer and ignores all those who don't or won't. Five people in every 400 typically develop this (very aggressive) cancer during their lifetime. If we take the figures at face value, the 20 per cent increase in risk if everyone – *every one of those 400 people* – eats an extra fry-up every day – and good luck to all who try – takes that toll from five cases to six. That is how we calculate a rise of 20 per cent. Twenty per cent of the original five is one additional case. We have gone from a 'relative risk', the 20 per cent, to a change in 'absolute risk' from 5 in 400 to 6 in 400, or 0.25 per cent. Relative risks make things look much more important.

Note that we have ignored – except by implication – what happens to the several hundred people in our sample who were unaffected before and are unaffected now, as did the *Daily Express*, as does just about every news report of cancer risk.

Let's change this. Let's begin instead by focusing on those who do not or will not have this cancer, the people for whom the revolver hangs idly on the wall and is never fired, those for whom the fire is never lit. Now we notice the 395 out of 400 who don't ordinarily get pancreatic cancer. If all 400 eat the extra fry-up every day for the rest of their lives, one more will have pancreatic cancer, but 394 will still be fine.

Now it looks less risky, don't you think? But we can make the numbers more striking by drawing attention to the fact that 399 out of 400 are unaffected by an almighty pig-out change in diet. There is, remember, only one extra case of this cancer, and the 399 were either going to get it anyway, or they were not and will be fine. Either way, 399 out of 400 are non-events as far as the extra fry-up goes. Yet this is exactly the same risk as the 20 per cent increase in risk if we concentrate only on the events and ignore the non-events.

The point is to ask if the way risk is usually talked about leaves the cup of danger pessimistically half-emptied by death, not half-full of life or survival. Except that in this case the cup is 399/400ths full of life, or at worst 394/400ths – but the newspapers look only at how much it empties.

Perhaps that's what people want when they talk about danger. We want to talk about the 1, not the 399. But this is a choice, even if we are unaware that we have one.

We could, as we say, talk about survival instead. And by concentrating

Figure 5: **If 400 people all have a fry-up every day, with an associated 20 per cent increased risk of pancreatic cancer, then one extra person (0.25 per cent) will get this cancer during their lifetime**

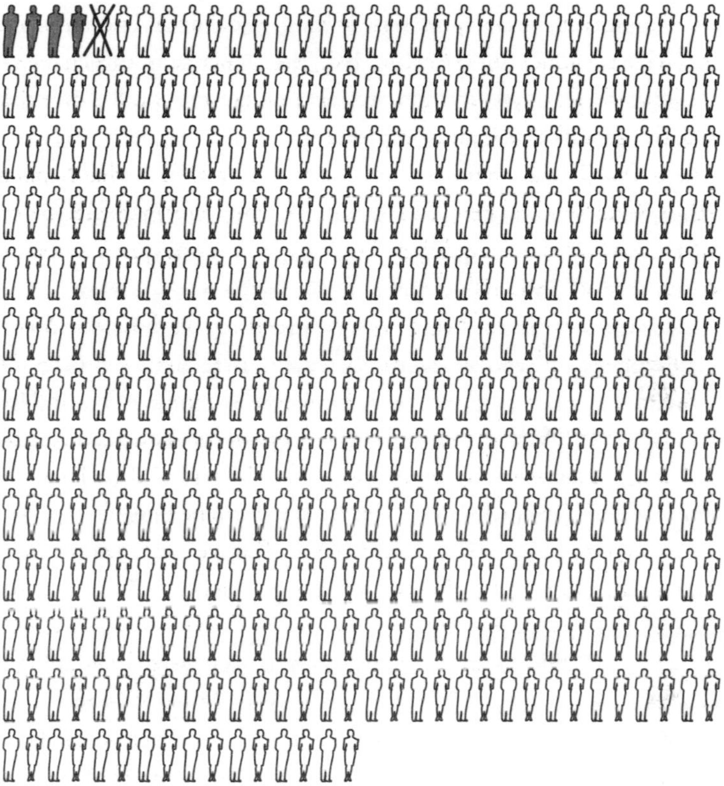

on those who are unaffected, the danger in this case diminishes. It isn't even our focus any more – living is – and the excellent prospect of survival is not much changed. We could also describe the pancreatic-cancer effect of an extra fry-up every day for life as roughly a 99.75 per cent chance of still being fine.

In Figure 5 above, is the risk of pancreatic cancer best measured by the people in grey, who get it, or by the people in white, who are fine? The extra chance of pancreatic cancer if all 400 eat a fry-up every day is shown by the person crossed out: is that best expressed as an increased risk of 20 per cent for the greys or a 0.25 per cent decrease in the chance of being fine for the whites, or as a 99.75 per cent chance of still being fine for the whites? It's all in the framing.

Although, following the work of psychologist Gerd Gigerenzer, there is a way to avoid percentages altogether, and talk about the pure number of people affected. Then, whether you want to talk about the increased risk, or the change in the chance of being fine, the measure is the same: one person in 400.

The man who ate an extra fry-up every day and didn't get pancreatic cancer and was fine is a massively more faithful rendition of how life is. It is also 'dull, dull, unbearably dull'. Newspapers never write it that way. People don't tend to talk that way. But then, the whole concept of risk assumes a 'negative frame': risk is about the potential for bad, seldom the potential for life to go on, except by implication, an implication that must lurk in the background. Standing in the foreground is death. Is it any surprise which perspective holds our attention? So if cancer is shown hanging on the wall in Act I, then cancer frames our expectations. Thinking about what makes the risk go 'up' tightens the focus further. We're mentally half-way to contracting it before we've even reached the last act.

Even MicroMorts do this. They too are about comparing the bad things that happen rather than the non-events. Perhaps they too need a complementary perspective, some sort of reframing. Maybe we need another new unit, an anti-MicroMort to describe the daily chance of being fine. How about a MicroNot (1MN) – a one-in-a-million chance that nothing fatal happens?

Then if we take as an illustration the average daily dose of acute fatal risk – 1 MicroMort – we can switch focus to the MicroNot to show that the average daily chance of an acute non-event – of being fine – is 999,999 MicroNots.

Then let's say that you do something, hypothetically, to double your average daily MicroMort risk from 1 MM to 2MMs – up 100 per cent. But reframe the risk with MicroNots to talk about non-events and your MicroNot dose falls from 999,999 to 999,998, a fall in your chance of being fine of 0.0001 per cent. Perspective, or framing, is all.

Another real news story illustrates a similar trick. Researchers found that a genetic variant – call it 'X' – present in 10 per cent of the population *protected* them against high blood pressure. Although published in a top scientific journal, the story received negligible press coverage until a knowing press officer rewrote the press release to say that a genetic variant – call it 'not having X' –had been discovered which *increased* the risk of high blood pressure in 90 per cent of people.[2] This story was widely reported, whereas talking about those to whom nothing happened wasn't news.

Both good and bad endings are present in any risk of less than 100 per cent; otherwise there is no risk, and the future is certain one way or the other for all. But the entire framing of risk is around the bad and changes to the bad, even when there is not much bad to be had, even when the good is overwhelmingly more likely and not much dented even by extreme changes in behaviour.

Sometimes, but rarely, the bad is changed to the good. There was an advertising campaign on London Underground that proudly declared: '99 per cent of young Londoners do not commit serious youth crime.' Sounds wonderful. But, taken literally, that means 1 per cent of young Londoners *do* commit serious crime, and there are around 1 million young Londoners, which implies 10,000 thugs running around. Not so wonderful. The switch between events and non-events, between fires that burn and fires that go out, can have a powerful effect on people's decisions. DS has found that when statins are shown to reduce the risk of heart attack for over-50s by 30 per cent, people are keen on statins. But they're not so keen when he also shows them that around 96 people in 100 will be unaffected if they take statins for ten years – they either would or would not have had a heart attack anyway – but might be vulnerable to side-effects. The risk hasn't changed, only the way it is represented; but that changes people's minds.

Does this make them irrational? For some, this is proof of human flakiness. Why aren't people consistent when the risk is the same?

We think that's unfair. A change in framing changes the context

– that's what framing is. People unsurprisingly find it hard to weigh a risk against everything in life that's relevant. Shown a risk in a way that emphasises danger, they react by saying it's dangerous. Shown it in another that emphasises being fine and they say: 'Oh, that changes things.' Our experience is that, once they have seen it both ways, they are more consistent. Some experiments that purport to prove irrationality through a change of framing seem more like a trick than a test. Do people who change their minds when the framing is changed change them back again if the framing changes back, and continue switching every time? Of course not.

In relation to stories, think of the numbers as follows: imagine 5 bad things out of 400 possibilities. This probability consists of a numerator and a denominator: 5 is the numerator and 400 the denominator. Narratives or stories often concentrate only on numerators. We tell stories about people who do things and people to whom stuff happens – the 5 – not the people who do nothing and to whom nothing happens. This is almost the definition of what stories are. If you want to know about a story you ask: 'What happens?' And so we suffer from what is known as denominator neglect. We ignore the mass of people from whom our 5 happening examples are plucked.

Not all stories are so simple, it is true. Good fiction teases out and tests our expectations and sometimes frustrates them. It puts revolvers on the wall and leaves them there. It can be full of ambiguity. In 1925 Virginia Woolf distinguished her fiction from that of the past when she wrote:

> If a writer were a free man and not a slave, if he could write what he chose, not what he must, if he could base his work upon his own feeling, and not upon convention, there would be no plot, no comedy, no tragedy, no love interest or catastrophe in the accepted style ... life is not a series of gig lamps symmetrically arranged.[3]

But this argument against orderly expectations has been used to exaggerate the modernist case. For it is also true that Hamlet's agonies about killing his uncle have exercised critics and audiences for 400 years with their lack of symmetrical, gig-lamp clarity, as does Hamlet's every relationship. There are no signposting coughs in Hamlet, the gig lamps are not bright and the play is brighter for it. Good fiction has played for centuries with causality and the limits of our knowledge.

But do the stories we tell in the news media or about our own lives have the same sophistication? Life is at least as puzzling and messy as fiction. But do we, many of us, behave as if it were a bad novel, and try to impose more direction and coherence than many writers would dare? Risk is not, or should not be, a bad novel. In real life the gun on the wall usually rusts. So maybe non-events should be bigger news. The *auteur* in Mitchell and Webb's sketch is that rare and uncelebrated thing: an artist of the denominator. Give him an Oscar.

All this is part of a general problem with risk perception known as availability bias, although, as we say, 'bias' is a mean word for what is often just a result of the framing. Availability bias refers to the fact that some things come to mind or attention more readily than others: the gun-on-the-wall effect. And sure enough, it's easier to think about events than non-events. Whatever it might mean to try to think about an event that doesn't happen, it is not easy or intuitive.

Researchers in the 1970s, among them the psychologists Daniel Kahneman, Amos Tversky and Paul Slovic, ran dozens of human experiments to discover what influenced people's estimation of risk.[4] They noticed that after a natural disaster people took out more insurance, then with time took out less. This was not because the risk rises immediately after a disaster then falls, obviously, but because the risk is more salient immediately after a disaster.

Slovic also found that tornadoes were seen as more frequent killers than asthma, although the latter caused 20 times more deaths. As we will also see in the chapter on crime, vivid events are recalled not merely more vividly but in the belief there are more of them. Put crudely, we worry more that something might get us not because it's more likely to get us but because it would make better telly.

This is a bias that can strike en masse, especially if we frame it right. The media play a part, and Kahneman uses the term 'availability cascade' to

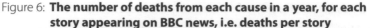

Figure 6: **The number of deaths from each cause in a year, for each story appearing on BBC news, i.e. deaths per story**

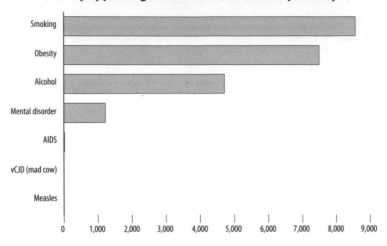

describe the surge of interest, attention to new cases and panic that come with a scary but rare event. In contrast, problems that are common are not surprising and are less likely to qualify as big 'news'. Another smoking death? And? So things that are genuinely likely to get you are not reported nearly so often as others that are rare. The unusual is, by the nature of news, disproportionately reported, so we think it more common.

Although this is a reporting bias, the media have no trouble justifying it on the grounds that people want to know about what's unusual and new, not what's old. There is no way they could report risk proportionately and still be in business. It would mean thousands of times more articles on smoking than on death from measles. But it is a bias, and although it is only speculation that this reporting bias affects people's estimation of the size of different risks, it seems like reasonable speculation. The extent of the bias has even been quantified – see Figure 6 above.[5] Mad cow disease, measles and AIDS received massive coverage relative to the number of people who died, whereas smoking, obesity and alcohol received little. The precise numbers of deaths attributable to all of these is debatable, and the fashion for certain stories has changed since the research was carried out more than a decade ago, but the thrust of the argument seems fair.

So there are all kinds of reasons why risk depends on what we pay attention to. This is partly because our attention is scatty, but also because it's not clear which version of the numbers is the one to watch: the numbers about what happens or the numbers about what doesn't.

5

ACCIDENTS

THEY WERE 11, Norm and Kelvin, the time they went to swim in the reservoir. It was warm as they pedalled along the lanes, and when they arrived the water looked cool. They threw down their bikes by the reeds, trod off their trainers, stripped to their pants – Norm's blue, Kelvin's white – and stood on the grass bank not far past the two fishermen. The dare should have been a breeze, to be honest, but Norm hadn't figured on the wind-up.

'No, really, they do!' said Kelvin. 'They're that long, with these evil teeth and they creep up underneath and go [snapping his jaws together] gnah!'

'Yeah, but not your bollocks,' said Norm.

Kelvin made rabbity, nibbling movements with his top teeth against his bottom lip, leaving brief pale slots in the pink skin.

'No way!' said Norm, looking away.

'Big. And pointed.'

Kelvin paused. Norm felt the breeze.

'Chew your knob off.'

'Shut up.'

'Here, pikey-pikey, Norm knob!'

'*Shut up!*'

For a moment there was only the water beating at the reservoir wall and the wind in the reeds.

'What's up?' said Kelvin. 'Cold?'

'Nah. You?'

'Nah.'

Norm half-folded his arms, then let them drop.

'So ... scared, then? Or what?'

'Me? Scared?'

'Go on then.'

The water chopped, and the waves slapped.

'Dive, yeah?'

'Yeah, 'course, dive.'

'You do it then.'

Danger isn't a choice at the age of 11; it's a game with rules. Kelvin stood on the edge looking at Norm over cupped hands. Then his arms flew back, his body was a breath and a twitch, his white legs flexed and leapt.

Kelvin was in, smashed by cold, then above the water, hitching up his pants, head high, blowing at choppy peaks across the reservoir. Norm twitched too. But his body refused. His feet held to the ground.

Norm picked up his clothes and held them to his chest, eyes on the dark head in the waves, his thoughts on the precious last inch of cold flesh in his underpants and the imagined snap and rip of a pike.

On the outside, Kelvin looked hard and fearless. On the bank, Norm was transfixed by a toothy, jelly-eyed monster like something he'd seen in a fish shop, staring at him close-up. As Kelvin pulled through the murk towards a soaked wooden platform on the far side, Norm cupped a hand between his legs.

Kelvin's swim to the platform would have been cold. Once there, climbing out seemed stupid as soon as he'd done it, except that Kelvin did it and Norm didn't. Then what? Swim back. Colder still.

As Kelvin swam his last, slower strokes, Norm looked down, one hand clutching his clothes, the other still down his underpants. Out of the corner of his eye he noticed the fishermen again. A shout, 'Oi!', turned into a clatter of wellies. Running.

Norm looked up: flapping boots, chopping water, running and shouting. He turned back to see Kelvin climb out, wherever that was.

He looked left then right, up and out over the water for the pale body and the dark hair.

'Kelvin?'

He twisted round. Looked back again for the place he'd missed. He saw water, green with flecks of grey and low, chopping waves, the platform, the bank, Kelvin's clothes.

'Jesus,' he mumbled and stared again.

'Kelvin!'

But there was no hair, no skin. Instead, his mind filled with the jelly eye and the sharp teeth, the eye, bite and blood. Fishermen plunged past into the water and the reeds hissed in the breeze.

He was white when they found him. But there was no blood. Still had bollocks too. The water did it, or maybe the cold. He wasn't far out, or very deep. Three more strokes, they said. He should have shouted, but the fishermen said he just went under. It wasn't Norm's fault, adults said.

Later, Norm felt vaguely disappointed that the big thing hadn't happened and his friend didn't die. Not even brain damage, they said. Then Norm felt vaguely guilty for feeling vaguely disappointed. At school on Monday it was cool to be Kelvin. Norm would have liked to be cool too, but he was still thinking about teeth and jelly eyes.

THERE ARE TWO WAYS to judge the risk of being bitten by a pike. One is 'Don't be daft, what pike?' Even if there are pike, they probably have zero appetite for Norm's willy. On the other hand, we could say the risk is stratospheric – because, come on, this is his willy he's thinking of. After that, what more needs to be said? The risk is beyond imagining.

This difference in these two calculations of the same risk is that between probability and consequence. Probability describes the chance of being bitten; consequence is the answer to the 'what if ...?' question. Is it likely he would have been bitten? Who knows? Probably not. But *what if* ...? Same risk, different perspective, sea-change in emotion and anxiety. Norm's struggle on the bank is between long odds and imagined agony. No prizes for guessing which grabs his attention. Just follow his left hand.

Fatal accidents are often like that: rare but terrible to think of. It is rare for children of Norm's age to drown: about 1 or 2 in every million children in the UK aged between 5 and 14 die through 'accidental drowning and submersion', each year, and about 3 or 4 in a million for those younger than 5. The figures are higher for males than for females at all ages.[1] There are no official figures for death by drowning by pike.

Even on the roads – where the threat of an accident is perhaps most apparent to parents (like us) who have clutched their children's hands – it is increasingly rare for children to die. But all this is little consolation if you think of consequences not probabilities. One in a million, as Prudence's mother said in Chapter 2, is useless to the one.

Take the true story of Mark McCullough, father of a seven-year-old girl who allowed her to walk the 20 metres from their home to the bus stop on the way to school each day, unaccompanied. Naturally, reported the *Daily Telegraph* in September 2010, he was threatened by a council with 'child protection issues'.[2]

The walk meant crossing a 'busy road' (said the council) or 'quiet country lane' (said Mr McCullough). The council was probably 'only doing its job'. Mr McCullough was 'very angry'. 'It's an absolute joke,' he said: 'I am not going to wrap my children up in cotton wool.' A fact confirmed one chilly morning when his daughter was spotted 'without a jumper', said the shocked Head of Transport Services.

As with drowning, the figures for the kind of accident the council fears are low. In 2008 in England and Wales there were 1,471,100 girls aged between 5 and 9. The Office for National Statistics (ONS) says 137 of them died from all causes. Seven died in transport accidents. One was a pedestrian.*

This is an average risk of less than 1 MicroMort per year. That is, 1/365th of the average daily risk of acute death for a UK adult. As we show below, more children die each year strangled by cords on window blinds.

Put aside arguments about parental responsibility and freedom. Concentrate on the risk – and answer an awkward question: which are most compelling, the basement-level mortality statistics, or the heightened fears of Lincolnshire County Council?

---

* In 2010 there were *no* pedestrian deaths in this category.

Figure 7: **Children killed and seriously injured on the roads, UK, 1979–2010[3]**

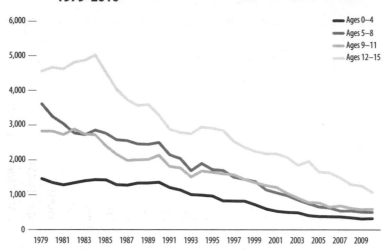

If you back the council, it may be because you feel the pull of what's called the 'asymmetry of regret', better known as 'How would you feel if …?' How would you feel if you took your child to and from the bus stop every day and your worst regret was that you wasted a few minutes to prevent an accident that was probably never going to happen anyway? Not entirely happy maybe, but in the scheme of things the time and effort are no big loss. Next, how would you feel if you did not give up those few minutes, then one sunny morning over toast and marmalade you heard the squeal of brakes?

Choosing what to do in life by trying to minimise what you might most regret is known in decision theory,[4] naturally enough, as mini-max regret. Most decisions are here and now. Imagined future regret is a tortured complication. It's more than the difference between short- and long-term self-interest. It twists the knife with potential guilt and blame, imagines this as if experienced with hindsight, then makes that anticipated-retrospective feeling our main motive, if you follow. It is a roundabout way of deciding what to do which people calculate in a blink. But at its most tyrannical, mini-max regret makes us slaves of nightmares. A bad ending foreseen or imagined comes to justify the

Figure 8: **MicroMorts per year for accidental injury, including transport, by age, in 2010, England and Wales**

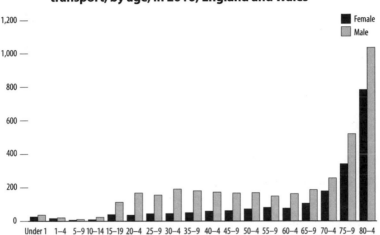

Equal to crude rates of death per 1 million population. Not including 'hospital misadventure', which we discuss in Chapter 23.

most drastic avoidance. Does it also save lives? Thomas Hardy said that fear is the mother of foresight. In the TV series *The Simpsons* 'just think of the children' is a joke that could be about mini-max regret. For some, it is a compulsion.

For people who feel this way, it is the potential 'what if?' – maybe the maximum imaginable 'what if?' – that they will do just about anything to minimise – that weighs far more than any odds. So it's no surprise that probabilities sometimes cut no ice, no matter how low, especially if the emotional stakes include children.

Still, let's remind ourselves – again – that the probabilities are lower than ever. As with mortality overall and also with murder, children are far less likely than any other age group to suffer accidental death.*

Old age, not childhood, is by far the most dangerous time for what the ONS calls avoidable accidental death.[5] The accident rate for over-85s

---

* We use two main ONS sources for accidental deaths: one defines 'avoidable' death and is derived from the other – a longer series – that simply counts all causes of mortality. The graphics here use the avoidable deaths series.

is so high that it would smash two and a half times through the top of this graph. We couldn't fit it on without making the risk to most other age groups vanish into the axis. For women, the clear trend is that death from accidental injury becomes more likely with age. For men, the trend is not quite so smooth, bucked in early adulthood and middle age by a taste for thrills that are bound to cause a few casualties.

If we highlight in these overall figures just the accidental deaths caused by transport, it's a slightly different story for young people. The old are still among the most vulnerable, but the 15–19s are now up among the highest.

Figure 9: **MicroMorts per year by age for transport accidents in England and Wales, 2010, equal to crude rates of death per 1 million population**

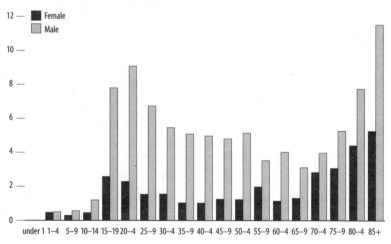

Note the different scale to Figure 8.

But as before, children aged under 15 are far less likely to die in a transport accident – which includes being hit by cars – than any other age group.

Many risks to children have fallen sharply over the last 50 years. You've probably met the carefree-youth bore with his 'I-used-to-roam-around-all-day-and-we-got-up-to-all-sorts-dodging-traffic/bullies/the-local-flasher-and-it-never-did-me-no-'arm'. Rose-tinted memories of childhood are a constant. Two out of two of the authors of this book might even have been caught expressing them. But they are often wrong. Many changes are clear-cut, and road accidents are a good example.

Watching a post-war Central Office of Information film, with men in hats and women in coats with big shoulders, reminds us how much roads have changed. In 1951 there were fewer than 4 million registered vehicles on the roads in Britain. They meandered the highways free of restrictions such as road-markings, traffic calming, MOT certificates for roadworthiness or low-impact bumpers. Children played in the streets and walked to school. The result was that 907 children under 15 were killed on the roads in 1951, including 707 pedestrians and 130 cyclists.[6] Even this was less than the 1,400 a year killed before the war.[7].

The carnage had dropped to 533 child deaths in 1995, to 124 in 2008, to 81 in 2009, and in 2010 to 55 – each a tragedy for the family, but still a staggering 90 per cent fall over 60 years. Over the same period the number of registered vehicles went up more than eight-fold, to 34 million.[8] If we assume a steady decline from around 1,000 deaths in 1951 to 50 now, this means an average of 450 children's lives have been saved in each of those 60 years: that is 27,000 people alive today who would have been killed on the roads as children if accidents continued as they had in 1951.

Of course, there have been huge improvements in emergency medical treatment that mean children in accidents, who in the past would have died, now run around to get knocked down another day. This is shown by the fact that injuries have dropped, but not as fast as fatalities: in 1951, 5,743 children under 15 were seriously injured on the roads, which went down to 2,502 by 2010, a fall of 56 per cent, though there is also much argument about the accuracy of the injury figures. One figure has not changed: around two out of every three child casualties are male.

But roads are not the only hazards for children. The ONS [9] reports that in total in 2010, out of 9.5 million children under the age of 14 in England and Wales, 246 died of non-natural causes, of which 172 were accidents. This included 21 pedestrians, 12 cyclists, 17 car passengers, 22 by drowning, 27 by accidental strangulation and 10 in fires.

This is an average for each child of 18 MicroMorts per year from accidents.

It is remarkable that about the same number of children were drowned as were knocked down as pedestrians. There were also more accidental strangulations than drownings or pedestrian deaths: looped window-blind cords are now recognised as a particular danger for toddlers around two years old, and in 2010 IKEA withdrew over 3 million blinds for this reason.[10] But if you are a parent, of which risk are you most afraid?

This huge reduction in accidental deaths and injuries must be good. Who wants more danger for children? But it is not always easy to know why children are not killed and injured on roads. Maybe a big reason is just that they are not outside walking to school or playing. So improvements may come at a price. We can report the numbers; we can't altogether explain them or their consequences. In 1971, about 80 per cent of seven- and eight-year-olds travelled to school without an adult. By 2006 the figure had dropped to 12 per cent of seven- to ten-year-olds, with getting on for half of all primary school children arriving by car. In 1971 the average age at which children were allowed to visit friends or shops on their own was seven. In 1990 this freedom was on average first granted to ten-year-olds.

These statistics reflect a growing risk-aversion on behalf of children, detailed by Tim Gill in his 2007 book *No Fear: Growing Up in a Risk Averse Society*.[11] Gill suggests a number of possible reasons for this: more cars, less open space, computer games, working parents and the greater publicity given to accidents and tragedy. Clichés such as the 'nanny state' and 'compensation culture' come easily to mind, but the issue is complicated.

Take playground safety. Some post-war playground apparatus now looks as if it was designed to injure: even children of the 1950s knew to watch themselves on the witch's hat and swingboat. In the 1970s, even before ideas of the nanny state and compensation culture took hold, the populist and popular TV programme *That's Life* ran a campaign for safer playgrounds, demanding new apparatus and re-surfacing. As Gill points out, these expensive changes led to a reduction in play provision, initially encouraging children to play in the streets, and had little benefit in terms of reduced injuries. This may have been through the phenomenon of risk compensation: DS saw this in his child's primary school when some large, ancient and much-loved piece of mouldering wooden playground equipment was replaced by a bright, new and rather dull climbing frame: in the first week a child tried to extract more excitement by balancing on the top bars, fell off and broke her arm.

Other countries have not been so obsessed with playground safety, and standards are now being relaxed rather than strengthened. The problem is that it is easy to quantify the danger of allowing, or even encouraging, children to have adventurous play, not so straightforward to quantify the benefit, which is hard to prove.

Gill argues that danger can be good for kids; they have an appetite for risk and will seek it out one way or another, so it is better to teach them how to manage risky situations in the future, such as from water or traffic, without getting into trouble. This benefits healthy development; it is, to use another cliché, character-building; it encourages a sense of adventure, entrepreneurism, self-reliance, resilience and all those other fine characteristics on which Britain built her Empire.

You often hear stories of people putting a stop to any activity that seems remotely fun by citing 'elf and safety' to avoid the charge of failing to look after children. Even the Health and Safety Executive thinks they go too far.

If that's not what you expect of the HSE, perhaps you've heard too much from its critics. But its public statements could have been written by Tim Gill. For example: 'No child will learn about risk if they are wrapped in cotton wool', sounds like Mr McCullough. The HSE goes on to say that it does not expect that all risks must be eliminated or continually reduced, or that every aspect of play provision 'must be set out in copious paperwork as part of a misguided security blanket'.[12]

We look more closely at health and safety risks in Chapter 18. The point here is that some low numbers for child accidents might indicate over-protection, which children pay for when they're older but no wiser, an argument the HSE appears to accept.

The chair of the HSE, Judith Hackitt, has said: 'Don't use health and safety law as a convenient scapegoat or we will challenge you. The creeping culture of risk-aversion and fear of litigation ... puts at risk our children's education and preparation for adult life.'

The Countryside Alliance found that in 138 local authorities in England, over a period of ten years, there were 364 legal claims following school trips, of which 156 were successful, resulting in an average total annual payout of £293 per authority between 1998 and 2008.[13] Compensation culture gone mad? In the five-year period between 2006 and 2010 the HSE brought only two prosecutions involving school trips.[14]

Occasionally the bad side of over-anxiety becomes clear. Obsessive use of suncream on children, or lack of outdoor activity in the sun, can lead to vitamin D deficiency, with the result that rickets, once thought to have been almost eliminated in the UK, is becoming an established disease of Middle England. Kellogg's even proposed to fortify its children's breakfast cereal with vitamin D, citing research which showed a 140 per cent increase between 2001 and 2009 in the number of British children under ten admitted to hospital with rickets.[15]

Tim Gill calls for a change from a philosophy of protection, in which every accident is seen as someone's failure, to a philosophy of resilience, meaning the ability to thrive in a world in which bad things happen.

As he says, 'Childhood includes some simple ingredients: frequent unregulated self-directed contact with people and places beyond the immediate spheres of family and school, and the chance to learn from their mistakes.'

Although children in the UK are more protected from accidents than ever before, the same cannot be said of all children everywhere. Road traffic accidents are the second most common cause of death of children aged between 5 and 14 worldwide, and the most common cause for 15- to 29-year-olds. An estimated 240,000 children a year are killed on the roads in developing countries. When we venture overseas, even those who encourage kids to walk to school in this country will shudder at the sight of small children in spotless uniforms, walking to school by the side of roads with thundering traffic and no protection.

6

VACCINATION

S HE SAW PRUDENCE turn again in the dark. Vile, black water streaked with acid poured across the girl's path. From nowhere, grey beasts with clanking teeth began to prowl while stinging insects steamed from the undergrowth, hungry. Nettles groped for her skin. Prudence hid her eyes and backed away – and stumbled into thorns that ripped her arms with poison as she fell to the forest floor where creepy-crawlies smelt blood and skittered over her legs to pierce her with pincers And as Prudence writhed and screamed, pricked, bitten and burned, her mother watched, lifted her foot and pressed a stiletto into her daughter's back to hold her down.

Everything a mother does is wrong, she thought, when she woke, disturbed, and later read that the dangers of life are infinite and among them is safety, according to Goethe, apparently. 'Thanks,' she said.

'Just a little scratch,' she said to Prudence after breakfast. 'Everyone has it. You can have a plaster.'

Because if safe could be dangerous, and the dangerous things were only allayed with more danger, which was really safe, and how risky was safety? Or whatever. Bloody Goethe. Bloody advice.

'It's to protect you,' she said, as Prudence sat in her TripTrap chair, colouring with little arms.

On the kitchen table lay printed pages from the Student Vaccine Liberation Army, 'armed with the most powerful knowledge' to encourage critical thinking about 'the risks and dangers of vaccines',[1] which explained how

> In England in the earlier centuries hatters used to use mercury to stiffen the felt of hats. The mercury was absorbed into their fingertips and they eventually went 'mad' ... MAD AS A HATTER. We are injecting mercury directly into our children's arms causing an epidemic of apparent madness ... ADHD, ADD, Autism, OCD, Bipolar ... are vaccine injuries mis-labelled as mental illness.

A leaflet from the doctor said:

> All medicines (including vaccines) are thoroughly tested to assess how safe and effective they are. After they have been licensed, the safety of vaccines continues to be monitored. Any rare side effects that are discovered can then be assessed further. All medicines can cause side effects, but vaccines are among the very safest.[2]

Bad if you do, bad if you don't, bad to Pru, bad to others. Preying on her mind were vaccination horror stories about children who died or suffered because they were *not* protected from common illnesses. 'If only we had given him the jab,' they said. If they had only vaccinated their own child, their friend's baby would not have been exposed to the virus before vaccination and would still be alive.

On her laptop was another mother's story: after jabs for hepatitis her baby screamed incessantly, hardly slept, didn't feed, woke one morning vomiting, was airlifted to hospital, survived with 'seizure disorder, cortical blindness, severe reflux and high risk for aspiration pneumonia ... severe developmental delay ... a mixture of hypotonia ... some spasticity', needing 24-hour care by two people.[3]

On the news last night was a report of a warning sent to schools about measles abroad, the risk of travelling if not immunised and the potential complications: eye and ear infections, pneumonia, seizures and, more rarely, encephalitis – swelling of the brain, brain damage and even death.[4] On top of everything else on the table was an appointment card.

In the end she went, telling herself to calm down but helpless with guilt, and for three whole days her heart beat at her conscience while she watched, and waited.

WHAT IF PRUDENCE had the jab and then soon afterwards began to show signs of autism? Is the story that her mum's decision to vaccinate was catastrophic? The truth is, we don't know exactly what causes autism – although there are plenty of theories. But when one thing follows another, the easiest story to tell is that one caused the other. So whatever she does, Prudence's mother feels she will be in the dock for whatever happens, as storytelling joins the dots in hindsight with any emotional glue, fear, politics, prejudice or presumption that come to hand to tell us that she got it wrong. Pity the mother, who can't win against risk when stories come to be told, after the event.

Pity the small boy too. When DS was a lad, there were no vaccinations against measles, mumps and chickenpox, so when anyone local had them he was marched round to be infected. The rule of thumb was that, if you were going to get it, it was best to get it over with now. Vaccination is a set of uncomfortable trade-offs, because DS now knows that measles exposed him to around 200 MicroMorts.[*] But there wasn't much choice back then, and no doubt it was character-forming.

Measles was common in the 1950s, the effects well known, complications not unusual. It can cause blindness and even death. Observe that in your neighbours' children and you cry out for protection. Because it was common, people saw it around, a clear and present danger they wanted rid of. Now that we don't see it so much, some cry instead for protection from vaccines. Visibility plays a big part in risk.

Given a choice between harm you can see and harm you can't, what would you do? Some avoid the one they can see – more threatening

---

[*] Between 1958 and 1960, 957,000 cases of measles were reported, of whom 178 died.

than a danger that's not here and now. Some take bigger fright at invisible risks, such as radiation, which seem more sinister. Others try for a cool-headed calculation of the pros and cons. And some of us simply do what the doctor tells us.

The simple point here is that 'risk' is more typically about 'risks', plural, and these risks often point in different directions: some immediate, some distant, some visible, some latent. How are you supposed to decide which risk to take when you are often judged not by the fact that you did your best in all good faith in a state of uncertainty but by how it turned out? Pru's mum sits at the table beset by tragic endings.

If she types 'vaccine safety' into her browser, she will see mostly official sites full of reassurance. If she types 'vaccine risk', she will find claims that children are harmed, that the science is bunk and the scientists are not to be trusted.

Vaccination ticks many of the fear factors that arouse strong emotions (see also Chapter 19, on radiation). Children are not even sick before we inject them. Actively sticking a needle in your child's arm is a sin of commission that hurts in ways that doing nothing doesn't – at least, not straight away.

Vaccination is also imposed, either through pressure or legal compulsion: if your child is to attend a kindergarten in Florida, for example, they must have been vaccinated against DTaP (diphtheria, tetanus and pertussis, or whooping cough), Hepatitis B, MMR (measles, mumps and rubella, or German measles), polio and varicella (chickenpox).[5] There can be side-effects. Finally, multinational corporations make a lot of money out of this mass-medicalisation.

All of which can provoke fierce opposition. And so claims that vaccination may cause dread outcomes such as autism find a ready audience, particularly in the US (although perhaps they compensate by not caring so much about GM foods).

At least children don't need to be vaccinated against smallpox any more, eradicated after the last natural case appeared in Somalia in 1977 (although a UK laboratory worker died in 1978). Smallpox killed countless numbers throughout history, with 2 million deaths a year right up to the 1950s. It helped Europeans conquer the Americas by almost wiping out the native populations. But it had long been observed that survivors did not contract the disease again, and the practice of *inoculation* developed, in which extracts from scabs were scratched onto the skin to give a deliberate but hopefully mild infection.

Then in 1796 Edward Jenner used another, much milder disease, in this case cowpox, after it was noticed that milkmaids tended not to have smallpox. Hence *vaccination*, from the Latin *vacca* ('cow'). Being scraped with pus taken from someone who had been exposed to an infected cow's udder isn't an obviously attractive idea, so naturally it was tried on a small boy – James Phipps, aged eight, the son of Jenner's gardener. Whether he gave informed consent is not recorded.

Moving vaccine around in the early days was tricky without refrigeration. The answer, as ever: small boys. To ferry cowpox to the Spanish Americas in 1803 – a voyage so long that one infected person would recover – eleven pairs of orphans were press-ganged into service and the first pair infected before they set sail. In time, they passed it on to the next pair, and so on for the duration of the voyage until good, fresh cowpox pus arrived in the New World. It took a while to catch on, but immunisation (which covers both inoculation and vaccination) went on to save millions of lives.

The history of measles offers a clue about the risks without immunisation. In England and Wales in 1940 there were 409,000 cases, of whom 857 died,[6] a 'case fatality rate' of 0.2 per cent or 2,000 Micro-Morts, the same as that quoted by the US Centers for Disease Control and Prevention (CDC) in the US.[7] Vaccination started in the 1960s, and by 1990 the number of cases had dropped to 13,300, with one fatality. Since 1992 there have been no childhood deaths from measles, only as adult consequences from early infection.

Vaccination can also be used to reduce long-term harm: it's been estimated that one cancer in every six is caused by infection,[8] but the long period between infection and illness make the association hard to spot. Nevertheless the HPV vaccine is now offered to 12-year-old girls to help prevent the infection that can lead to cervical cancer.

So it seems a good thing to be vaccinated. Rather like stopping smoking, it is also good for people around you. This is because of herd immunity, when sufficient numbers of people are immune for an infection not to turn into an epidemic. What 'sufficient' means depends, in a fairly simple way, on how infectious the disease is. For example, a single person with smallpox would infect, in a community of susceptible people such as the Incas, on average around 5 others.* If they in turn infect 5 each, and so on, then it will only take six steps to infect an entire community of 50,000 people.

Given the infection rate for measles, we need to vaccinate 92 per cent of the population to prevent the spread of an epidemic.† The rate in three years of the mid-2000s was 82 per cent, 80 per cent and 81 per cent. By 2011 it had recovered to 89 per cent. Figure 10 (overleaf) shows what happened for cases of measles in England and Wales.

---

* The average number of people infected is called the *basic reproduction number* and given the symbol Ro (R-nought) – for smallpox it is around 5, for measles around 12.

† Suppose the community had been vaccinated, to the extent that more than 4 out of 5 people (80 per cent) were immune. In this case, someone with smallpox would on average only infect less than 1 new person, and so the epidemic would die out. So we get the nice general result that an epidemic will die out provided that of every Ro people, at least (Ro – 1) are immune, so the proportion that need to be immune is (Ro – 1)/Ro. For example, to prevent a measles epidemic (12 – 1)/12 = 11/12 = 92 per cent of the population need to be immune, whereas the swine flu virus in 2009 was a lazy thing whose Ro was only around 1.3, and so only 0.3/1.3 = 23 per cent had to be immune to stop the epidemic. Since vaccines are not 100 per cent effective, in practice the aim is to vaccinate a greater proportion than this, at least 95 per cent in the case of measles. Of course, it's possible to get a 'free ride' by not having the vaccination yourself, and relying on everyone else to stop an epidemic.

The current (2011) English vaccination rates for measles[9] are 89 per cent, up from 80 per cent in 2003 but still not back to the 92 per cent in 1995, let alone the 95 per cent recommended by the World Health Organisation (WHO). After an outbreak of measles in Liverpool in February 2012, the Health Protection Agency revealed that 7,000 Merseyside children under the age of 5 had not had their full measles vaccine. Measles is the first M in the MMR vaccination, and coverage went down after the highly publicised claim in 1998 that MMR was associated with autism. This claim has now been discredited, although it continues to have strong supporters in the US – just try searching on 'vaccine autism'.

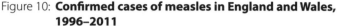

Figure 10: **Confirmed cases of measles in England and Wales, 1996–2011**

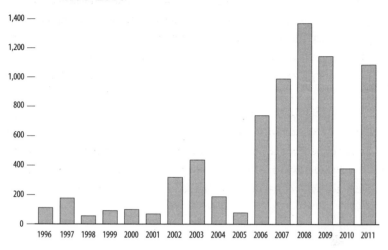

The fact that any one of us can have most of the benefit of vaccination without actually being vaccinated is known as a 'free-rider' problem. You can rely on other people to keep the risk low – as long as they don't stop too.

So, is it risky not to be vaccinated? It is, and it isn't. The risk could be nil, and it could be huge. That is because it depends on what other people do, as well as what you do. If they all carry on vaccinating and you stop, you'll probably be fine. If they stop too, you could be in trouble. The risk is dynamic and contingent. We are both subject to the risk – from others – and we are the risk, because we could become infectious.

So exactly the same behaviour on our part can mean extreme differences of risk, depending on what other people do. This makes reliable numbers about the risks that you face as an individual impossible to calculate. There's a big risk down the road for all of us if there is a failure of herd immunity, but what that means for you, now, is impossible to say. You could be fine. But what if you contribute to a loss of herd immunity? You could be dead. So, what's the risk?

But there is no denying that vaccines can have side-effects. For example, the UK Medicines and Healthcare Products Regulatory Agency

(MHRA) publishes the adverse event reports received following at least 4 million doses of the HPV Cervarix vaccine.[10] There were 4,445 reports listing 9,673 reactions, although these reports are voluntary, and so the rough rate of 1 report per 1,000 injections is a major underestimate. The US CDC warns of mild to moderate reactions in 1 in 2 cases.[11] Most of the MHRA reports are minor consequences of the injection such as pain and rash, and over 2,000 were considered 'psychogenic', caused by the injection process rather than the vaccine itself, including dizziness, blurred vision and cold sweats.

The problem comes with rare but severe events that occur later, and it is debatable whether these were due to the vaccine or would have occurred anyway. The MHRA lists over 1,000 reported reactions that they do not recognise as associated with the HPV vaccination, including four cases of chronic fatigue syndrome. Given the number of 12- to 13-year-old girls in the programme, the MHRA estimated they would expect to see 100 new cases of chronic fatigue syndrome over this period anyway, regardless of vaccination, so it is remarkable how *few* cases have been reported. But those families may well be convinced that the vaccine caused their child's condition. It is the 'available' thing to blame.

The real problem is that with any mass intervention there will always be bad occurrences around the time of the jab – essentially, coincidences. For example, in September 2009 the *Daily Mail*'s headline declared that 'Schoolgirl, 14, dies after cervical cancer jab'[12] and quoted the head teacher as saying, 'During the session an unfortunate incident occurred and one of the girls suffered a rare, but extreme reaction to the vaccine.' Three days later it was revealed that the girl had cancer and the death was coincidental:[13] however this was not headline news, and this tragic event is used repeatedly on websites as proof of the dangers of the HPV vaccine.

But sometimes the reports are real. There was a classic example in 1976, when a new strain of swine flu was identified in Fort Dix, New Jersey. Fearful of a repeat of the 1918 flu epidemic that killed millions around the world, the authorities ordered mass vaccination, and 45 million people were immunised.

Within a year the programme was abandoned, for two reasons. First, around 50 cases of Guillain-Barré syndrome – a gradual paralysis that is now thought to have been Franklin D. Roosevelt's condition[14] – were reported, and 500 were eventually identified among those vaccinated. This suggested that among every million who had the vaccination, 10 more people than usual would get Guillain-Barré.[15] In all, 25 died.

The second reason the programme was stopped was that the epidemic never got out of Fort Dix: nobody else had the flu, and there seemed no benefit to balance the possible harm. The Director of the CDC was later sacked but still believes the vaccination programme was right.[16]

Not all flu vaccines have the same risks. Following the UK swine flu outbreak in 2009, 9 cases of Guillain-Barré syndrome were diagnosed within six weeks of vaccination, but it was concluded this would be expected by chance alone.[17] But Finland and Sweden have reported increased rates of narcolepsy – sudden paralysis and sleepiness – in children after the swine flu vaccination, and this is still being investigated.

As the MMR saga showed, it's hard to disprove an association. Thimerosal is a preservative used in some vaccines and contains mercury. It has long been accused of harming children. The CDC say there is 'no convincing evidence of harm', but it was agreed in 1999 that it should be 'reduced or eliminated in vaccines as a precautionary measure'.[18]

The official line that the overall benefits of vaccination outweigh any risks ignores the way in which imposed and highly visible harms, however rare, are seen very differently from potential benefits some time in the future, which can never be confirmed and seem 'virtual' in societies where the risks of infectious diseases are so low.

It is a different matter in less-developed societies: the WHO report that there are still 140,000 deaths from measles each year, one every four minutes.[19] And, as we have seen in England, these are preventable. Vaccination has already made huge inroads: there used to be 2,600,000 deaths a year. The eradication of measles is thought to be feasible, just like smallpox, especially now the vaccine is stored in the fridge, not in a small boy.

COINCIDENCE

FOG EVERYWHERE. Fog lying in city streets like sleeping vipers. Fog, ghostly and concealing. The fog of novels, thick, bleak and secretive. The fog of spies, mysteries and thieves.

Norm, 18 years old, walking. Hesitant in the gloom. Lost in thought of his father, who set sail two years ago to the day in his adored sloop *Bill*, for a sally off the coast from Lymington to Cowes – as was often his habit – and in the most perfect of English zephyrs disappeared, boat and all, never to be seen again. How Norm yearned to hear once more the old man's rambling seafarer yarns. Even the fog seemed to whisper memories of the sea.

As he stole through the damp, grey air, Norm's left foot chanced to strike upon something soft. He was of a mind to walk on but fancied he heard a moan, and the humour came upon him to investigate, whereupon he discerned through the fog a human form, face down on a park bench, legs sprawled across the path. He knelt and examined what turned out to be an elderly man. To Norm's astonishment, it was the silent but unmistakable and thankfully still breathing figure ... of his very own father.

Over a restorative cup of hot chocolate at a nearby Costa, while outside the fog still crept, the bearded man's remarkable story unfolded, of sudden loss of memory, of drifting vacantly to what he later discovered was France, where he woke in only a pair of shorts and sandals, without identity; of months wandering, picking up casual work as he could, avoiding the authorities for fear that he carried with him some dark secret about poisoned fish, of how he fell in with a gang of pickpockets who roamed Paris and only yesterday had come to try their criminal luck, as luck would have it, in Basingstoke, and how, as hazy recollections began to return to him in this his home town, betrayal and lies saw him abandoned late in the evening, asleep on a park bench.

'I think,' he said on the slow bus home, trying to understand his sudden state of bewilderment those two years ago, 'it was the shock of our lottery win – £100,000 it was – and then my silly superstition about banks such that I chose to bring it home in cash in a Tesco carrier bag, then left it all on the number 63.

'After so many years struggling to raise money for my old orphanage in Clacton, threatened by that developer, the loss was too much to bear. Out at sea, my poor, tortured soul and wounded mind simply chose to forget ... everything.'

'You never know, father,' said Norm, 'it might turn up. Who knows, it could be lying undiscovered under this very seat,' he laughed.

And Norm playfully reached down. There was indeed an abandoned plastic Tesco carrier bag wedged in the frame under the seat. Did they not clean buses nowadays? He pulled it out and peered inside at the usual lumpy detritus people were in the shameful habit of casting hither and thither, only to discover, amazed, great piles of tightly bundled £50 banknotes.

'My God,' said Norm's father, 'that's it! I'd recognise the money anywhere.'

'Excuse me, gentlemen,' said an elderly female voice behind them. They turned to see a small, frail woman in a black shawl, peering strangely at Norm's father.

'I couldn't help noticing the unusual birthmark on your right ear. And then when you mentioned the orphanage at Clacton, a shiver ran through me. Forgive an old lady, but I've been searching for so long, you see, and I fear my mind plays tricks upon me. But I believe ... no, one more question before I can be certain: did you ever wear a silver chain bearing the small figure of a cat?'

'Ah' said Norm's father, 'I'm afraid I did not. I perceive your hopes, but I am not the son you seek.'

The old lady sank visibly. Another hopeful trail, perhaps the most hopeful of all, had led nowhere.

'But I know who is …'

She looked up. She smiled, hesitantly, a hopeful light restored to her eyes. She could scarcely contain the beating of her frail old heart. Could it be true?

'For there's someone else we must both now find together, an old friend, Bill, my greatest friend, a friendship first fashioned in our days in the orphanage, joined as we were in an improbable bond by identical birthmarks on our right ears. He always wore a silver chain exactly as you describe. He suffered a terrible breakdown in mid-life at the memory of his abandonment in childhood and was confined thereafter to an institution. But alas, my own memory fails me now, and I cannot recall where. And yet for some reason I picture the sign of a ship in harbour.'

'Like that one?' said Norm, looking out of the bus window at the very moment when, by chance, the fog began to lift, and pointing across the road to a sign above the door of a house of tranquil and fetching beauty.

'My God, that's it!' exclaimed Norm's father, by remarkable coincidence for the second time that day.

A few moments later Norm and his father were witness to the most tearful but blissful of reunions. At the sight of both his mother and his old friend, Bill recovered his wits in an instant. Moreover, the old nurse who had tended him with such devotion beyond her retirement age on account of the striking resemblance of a birthmark on his right ear to that of the son she too had been forced by cruel luck to give up to an orphanage in Clacton, was, of course, the mother of Norm's father, Norm's grandmother, only that day reeling from the news that the orphanage was finally to close for want of a last £100,000 – and with it the last connection of many a mother to her long-lost child.

'Perhaps, Granny,' said Norm, 'this will help.' And he held forth the carrier bag.

'You know,' said Norm's father, 'it reminds me of the time I was a-going around the Cape in the blackness of night in a force 9. The sea 'twas cruel, the ship's very hull did answer the wind in pain ...'

ONE DAY, on a cycling holiday stop in the Pyrenees, Mick Preston set off to the post office with a postcard for his mate Alan. On the way, who should he meet coming up the street, but Alan, on holiday. So he handed him the postcard.[1] As Mick said: a waste of a stamp.

What coincidence? Oh, that coincidence. Fancy meeting the very person, etc. Well, if you say so. But how much of a surprise is it? Call it spooky, weird, whatever, but is it really so improbable that this should happen to someone, somewhere, at least once, and since it happened to Mick, is it a surprise that he talks about it? Events bump into each other all the time; it is only when people notice that we use the subjective word 'coincidence'.

In *The Art of Fiction* the author and critic David Lodge says: 'Coincidence, which surprises us in real life with symmetries we don't expect to find there, is all too obviously a structural device in fiction.'[2]

If true, that says something odd about us. Because this chapter is full of coincidences, and they are all real. If we don't expect them in real life, well, we should. There are plenty about. Though maybe not quite so many per head as Norm experiences one foggy day in Basingstoke. So is one reason that we often find coincidence clunking in fiction – as David Lodge says – because we make too much of it in life? Is coincidence over-rated all round?

DS invited people to write to his website with their coincidence stories,[3] a number of which follow. He received thousands, all in one way fabulously unlikely. But if fabulously unlikely things happen so often, can they be that unlikely?

The problem – though it's also a huge evolutionary advantage – is that we are born seekers of cause and effect – as with the links between vaccinations and reactions – and we can't help trying to work out why things happen. 'Nothing happens without a reason' is almost a description of how our basic cognitive kit functions. So if there is no particular reason, we're easily foxed. To call coincidences random is almost two fingers to our normal sense of control and meaning.

But coincidences of the 'fancy-meeting-you-here' type are certain, if not for you then for someone, somewhere, and they needn't mean anything. The opportunities are infinite, so why the fuss? There are enough of us with enough possible connections to make even the ingenious plot-driven inventions of Charles Dickens just about plausible. These coincidences are sometimes called artistic licence. Do they need any? Or do they just reflect the fact that when lots of things happen, as things do, some are bound to be strangely coincidental?

But let's not strip all the romance from coincidences by making them trivial just yet. Mostly, so far, we have been dealing with the dark underbelly of risk – accident, death, disaster, gloom and doom – but coincidence can show the bright side of the play of chance in our lives. That's why coincidence has an important place in a book about danger, as a version of danger's opposite – the turn-up.

What is a coincidence? It's been defined as a 'surprising concurrence of events, perceived as meaningfully related, with no apparent causal connection'.[4] These concurrences may be two things that happen at exactly the same time: for example, a parent and child whose letters to each other cross in the post – after 37 years without contact.[5]

Almost everyone seems to have a story of meeting a familiar figure in some unexpected place, or discovering some unexpected extra connection, such as the engaged couple who found they had been born in the same bed.[6] Objects feature too: such as buying a second-hand picture frame in Zurich, and finding in its lining a 30-year-old newspaper cutting containing your own photograph as a child,[7] or being on holiday in Portugal and finding a coat-hanger that belonged to your brother 40 years previously.[8]

Why do these extraordinary events happen? Various strange forces have been invoked, such as Paul Kammerer's principle of 'seriality': 'The central idea is that, side by side with the causality of classic physics, there exists a second basic principle in the universe which tends towards unity; a force of attraction comparable to universal gravity'.[9] Kammerer says seriality is a physical force, but he dismisses as superstition any supernatural ideas that could, for example, link dreams to future events. In contrast, the psychoanalyst Carl Gustav Jung revelled in paranormal ideas such as telepathy, collective unconscious and extra-sensory perception, and coined the term 'synchronicity' as a kind of mystical 'a-causal connecting principle' that explains not only physical coincidences but also premonitions.

Alas, more mundane explanations are possible.[10] First, some kind of hidden cause or common factor could be present – maybe you both heard that the Pyrenees was a nice holiday spot. Psychological studies have identified our unconscious capacity for heightened perception of recently heard words or phrases, so that we notice when something on our mind immediately comes up in a song on the radio. And of course, we only hear about the matches that occur, and not of all the people you have spoken to with whom you had nothing in common and indeed were pleased to escape. Few feel excited enough by not meeting a friend in the Pyrenees to tell anyone about their non-event.

Even so, people find coincidences weird. Let's try another way of making them ordinary – by making the extraordinary common. When DS tried his luck on the National Lottery with the numbers 2, 12, 15, 25, 32 and 47, he lost. The winning numbers that came up that week were 4, 15, 19, 44, 45 and 49.

Extraordinary!

'What's extraordinary about that?' you say. He lost. Yes, but what are the chances of his six and the winning six being in this combination of 12 numbers? This is an amazingly rare combination of circumstances with a probability of 1 in 200,000,000,000,000 (1 in 200 trillion), the same chance of flipping a fair coin 48 times and it coming up heads every time. Impressed?

No, you're not impressed, we can tell. But why not? Well, you say, because the lottery had to come up with some set of numbers and so too did DS, and what's clever about them being different? But that misses the point. The chance of that precise combination – or of any other – was vanishingly small, *and the same as any other*. And yet it happened, as some combination has to happen. Only after the event does it seem boring, and only then because we give meaning to it. But the numbers don't care about the meanings we attach to them. They just come up with the same massive degree of unlikeliness in every case. But again the question, if everything is equally unlikely ...

So it's purely the problem of predictability by people that gives this week's lottery numbers any meaning. The probabilities, or rather the improbabilities, remain the same for all combinations of guesses and results.

The chance of winning the jackpot on Britain's National Lottery is about 1 in 14 million. The only thing you can do to improve your prospects is try to pick numbers that other people won't, so you're less likely to have to share your winnings. So, since many people pick birth dates, avoid numbers below 31 (no one is born on the 32nd of the month).

Something similar applies to coincidences. Predicting precisely which coincidence will happen would be next to impossible. But afterwards: so some numbers/coincidences came up? And ...? Some combination of events is inevitable. Things have to happen. Maybe what happens is not meeting a friend in the Pyrenees, but maybe the Alps, or maybe not a friend but a relative, or maybe not carrying a postcard but having thought of them at breakfast, or maybe while singing their favourite song, or maybe not on the path but in the bar, or maybe not to this person but to someone else, or maybe ... suggesting that there are innumerable potential coincidences and presumably innumerable near-misses. And whatever it is that happens to happen among all these possibilities that also turns out to be memorable, that's what happens to be talked about.

Coincidences are just an excuse to pretend that big numbers with limitless possibilities are meaningful in the small and everyday compass of our own lives. They're not. They just happen to catch our attention. They are a human vanity.

But tell that to the passenger who travelled by coach from Limerick to London and took a copy of *The Name of the Rose* to pass the time, then left the book, by accident, unfinished, on the bus on arrival in London. On the return trip about three months later, there in the pouch at the back of the seat, was a different copy (different cover) of the same book.

Simple chance can be a strange and unintuitive force that throws up surprising concurrences more often than we might think, since truly random events also tend to cluster. Just as, if you throw a bucket of balls on the floor they won't arrange themselves in a regular pattern but will probably cluster here and there, so people moving randomly will some-times find themselves gathered in one place. Ever been sitting on the Tube to look up and find all the passengers on one side?

This characteristic produces brain-mangling results. For example, it famously takes only 23 people in a room to make it more likely than not for two to have the same birthday. We think the best quick way to see this – for those who find it intuitively outrageous – is simply to get a feel for the number of possible pairs of two people from 23 people. Say at the end of a game of football every one of the 23 people on the pitch (two teams and a referee) had to shake hands with everyone else, there would be 253 handshakes (and thus 253 possible birthday pairs). That is, there are an awful lot of possible combinations of two from among 23 people.

This, of course, means that in around half of all football games there will be two people on the pitch who share a birthday. Maybe they could give each other a hug. And that's just birthdays. Now let's think of every-thing else that any two players or the ref might have in common.

Given all the possible places and all the possible ways that two acquainted people could meet, or all the possible things that appar-ent strangers might have in common, once you start multiplying the possibilities you wonder if every day on the seats of London Transport or slumped on a park bench in the fog, or tucked in a discarded Tesco carrier bag there sit more potential connections than particles in the universe. Some will be a match, somehow, and never know it. Think of Dickens as choosing his plots from an infinity of real-life events, after the event: perfect realism. The only reason Norm passed the audition to be in this book is because he had experienced this remarkable set of coincidences. Otherwise, we'd have chosen someone else.

The final explanation for coincidences is what is called the 'law of truly large numbers', one version of which says that anything remotely possible will, if we wait long enough, eventually happen. So even genu-inely rare events will occur, given enough possibilities. Take three chil-dren in a family. The first comes along on a certain day. The chance of the next one sharing the same birthday is 1 in 365, multiplied by another 1 in 365 chance for the one after that, which equals a 1 in 135,000 chance of them all sharing the same birthday, and better if there is planning going on. This is rare. But there are a million families in the UK with three children under 18, and so we should expect around eight of them to have children with matching birthdays, with new cases cropping up around once a year, which they do: new examples in the UK occurred on 29 January 2008,[11] 5 February 2010[12] and 7 October 2010.[13] And these are only the ones that got in the papers.

It would be truly strange if memorable coincidences didn't happen to you. But this may be difficult to keep in mind when you're walking past a phone box, it rings, you decide to answer it, and you find the call is for you.[14]

It's the detail that makes us sit up. 'A postcard, you say? Fancy that. In the Pyrenees? Well, of all places.' But the more detail we allow, the more possible points of contact there are. Just think of all the people you have ever known. Then think of all the people that you have had some connection with, such as attending the same school, being friends of friends or family and so on. There will be tens of thousands. Perhaps the best way to think about coincidences is not how rare and strange they are, but how many we might be missing. If you're the sort of person who talks to strangers, you will find more of these connections. If you're not, then you have probably sat on a train next to a long-lost twin from whom you were separated at birth and never realised it. Coincidences are simply a refreshing reminder that we live among infinite possibilities. Even as that thought makes them less surprising, it also makes them wonderful.*

---

* There is some nice, fairly simple maths that allows you to work out how many people you need for a good chance of a match for any characteristic.[15] Suppose that any two people have a 1 in C chance of matching – for example, for an exact birthday match, $C = 365$. Then to have a 50 per cent chance of a match in a group of N people, N needs to be around $1.2 \sqrt{C}$. For a birthday match, this means that we need around $1.2 \sqrt{365} = 23$ people, as claimed above. For a 95 per cent chance of a match, we need to approximately double this number, to $2.5 \sqrt{C}$. So if we have $N = 2.5 \sqrt{365} = 48$ people in a room, it is very likely indeed that two will have the same birthday.

This makes it easy to make money off people. Suppose you have 30 people

That might never free coincidences in fiction from being obvious structural devices, but perhaps we should go easier on them. The only problem is, if we did lose some of our wonder at coincidence, would we lose some of our fascination for the story?

---

together. Bet the group that two have a birthday within one day of each other. What are the chances you will win? First, consider the chance that any two people (say Me and You) match in this way: if My birthday is 16 August, then a match would happen if You were born on the 15th, 16th or 17th, which is 3 out of 365 days, or a 1 in 122 chance, so $C = 122$. So for a 50 per cent chance of a match we only need $1.2 \sqrt{122} = 13$ people, and for a 95 per cent chance we need $2.5 \sqrt{122} = 28$ people. So with 30 people in a room you are almost certain to win. You might send us a cut of your winnings.

8

SEX

KELVIN'S DIARY, AGED 19¾:

Woke up. Head hurt.
Kath in the bed – asleep. 'Oi, Kath!' Had shag.
Read paper. Seemed familiar.
Last week's paper.
Opened curtains. Sunny day.
Closed curtains.
Half tin Stella under bedside table. Downed Stella. Had shag.
Slept.
Woke up. Hungry.
3pm. Found socks. Off to Mr Singh's.
Bought lunch for two: tin of Fray Bentos Ready-Made steak
    and kidney pie, 2 Flakes, tin of Stella.
Ate Flake. Ate Kath's Flake. Drank Stella.
Home. Hi Kath.
Put tin of Fray Bentos Ready-Made s&k in oven.
25 minutes. Long time. Turned up oven.
Had shag (sofa).
Had kip (sofa).
Got up.
Strange smell. Brushed teeth.
Strange smell cont'd.

Remembered tin of Fray Bentos Ready Made.

Took tin of Fray Bentos Ready Made out of oven.

Tin of FBRM very large/spherical. Also black + red glow.

Realised had forgotten to pierce tin of Fray Bentos Ready-Made before cooking as instructed.

Pierced tin of Fray Bentos Ready Made.

Whooshing sound.

Tin of FBRM very fast.

ToFBRM flight-path erratic.

Ducked.

ToFBRM hit cupboard.

Kath not ducked.

Kath struck on shoulder by rebounding tin of flying red-hot FBRM.

FBRM does spinning thing on floor until hiss gone.

Kath lying down, moaning.

Can see knickers. Red. Looks nice.

Suggest shag.

No shag.

Hospital crowded.

10pm. Home. No Kath. Kath parents collected Kath, Kath fractured clavicle.

Eat tin of FBRM. Nice.

Call Emma.

BE HONEST, WHAT GRABS you: the number 5.6 per 100,000 population or a pressurised tin of Fray Bentos ready-made steak and kidney pie pinging around the kitchen? The first is the rate of new cases of syphilis diagnosed in the UK in 2011. The Fray Bentos is an image, no more, a crazy metaphor for carelessness. But it's the image that sticks in the mind.

Visual images of danger are usually sexier than numbers, for obvious reasons: they smack the senses with sound, colour, movement and violence. More sneakily, they also slant the way that the danger appears, showing us its consequences, often ignoring the probabilities.

It's the same split between consequence and probability that we met in the chapter on accidents, when Norm fixated on losing his willy to a pike. It's not likely, but it's everything. Pictures of danger are usually the same, a vivid image of a bad 'what if?'*, not a probability. The danger of skiing – which, based on the odds, ought to be illustrated by someone simply skiing, safely, having (dare we say it?) fun – is more likely to be pictured as a leg in plaster or a whirl of skis and snow off a precipice. We don't 'see' odds – how likely the thing is – we 'see' consequences. That's what people would mean if they were to say 'picture the risk'. They mean picture the worst that can happen.

Advertisers and governments know this. When they want to change our behaviour because 'it's risky', or persuade us to buy something to make us safer, they often choose images of worst-case 'what if?'s.

One of the most famous sex-risk images in the UK was what's known as the AIDS monolith advertisement in 1987. As the smoke clears, we see something like a tombstone carved in rock blasted from the mountainside. The chilling, brilliant voiceover is by the actor John Hurt:

> *There is now a danger that has become a threat to us all.*
> *It is a deadly disease and there is no known cure.*
> *The virus can be passed during sexual intercourse with an infected*
> *person. Anyone can get it, man or woman.*
> *So far it's been confined to small groups, but it's spreading ... If*
> *you ignore AIDS, it could be the death of you.*

Other public information films about danger from a time when governments still made them include one about road safety with an image of a hammer swinging into a peach:

> *It can happen anywhere, to anyone.*
> *An ordinary street.*
> *A moment's thoughtlessness.*

Then splat, the hammer hits the peach.

---

* The fashion for data visualisation is an exception, where pictures of the numbers can be striking. See, for example, David McCandless's *Information is Beautiful*.

Another, on the risk of crime, has hyenas prowling around a parked car. All very vivid and scary – and all consequential, not probabilistic.* But then, how do you do an image of one in a thousand, or any other odds? Not easily, not vividly. Numbers don't splat like the flesh of a peach or a human being; they don't bite like hyenas. In contrast, images of the worst-case 'what if?' are easy, especially for one victim. And then comes the extrapolation to everyone else, as we're told that this tombstone/ hyena/splatted peach could be you. Think again of Norm's paralysis as he imagines the teeth and glassy eye of the pike in the reservoir. The consequential image dominates the odds for him too.

Better than public information films and their selective images – don't you think? – is private information. 'I know from personal experience ...' are the kind of words we use with all the plumped-up authority we've got, certainly more authority than we grant governments. But personal experience and sex? Authoritative and rigorously unbiased? Not selective? Yeah, sure.

Let's say you have unprotected sex, like Kelvin and Kath. If nothing goes wrong, you feel relieved. But you might also feel the risk probably wasn't that high after all. Exposure to danger can have that effect – of making people feel safer. 'See, it was OK!' One formal explanation for this is that small samples of rare events are skewed. This means that when bad results are not typical – say only one time in every 20 – you might get away with the first few so you expect to be fine next time too. Your chance of being all right is 19 out of 20, or 95 per cent.

In truth, you shouldn't deduce anything about the true odds from one experience. But deduce you will. Based on your own massively selective sample of the past, you might learn to underestimate probabilities in the future. 'Had shag, no kid', as Kelvin might say. Do this a few times and he will – probably – still be OK, encouraging him to think – as if he needs encouragement – that this is the sort of risk he can take.

On the other hand, if it all goes wrong, people tend to overestimate the risk that it will go wrong again. Suppose unsafe sex leads to an STI

---

* On the National Archives you can watch the AIDS monolith film, the peach and hammer (from 1976) and the hyenas.[1]

about one time in 50. It doesn't, but bear with us. People might not know the figure, even roughly. After 100 people have five encounters each, roughly ten have an STI.

These ten might say, 'Crikey, we did it only five times and look what happened.' So they take away an exaggerated sense of the odds. The other 90 might imagine that the probability of danger is not much above zero: 'Well, it never happened to me.' There are lab experiments – not involving sex – that seem to confirm this pattern of belief and behaviour.

So personal experience can lead us astray. Again, we need bigger data than just our own notches on the bedpost.

Sex, like so many other fun activities, clearly carries dangers. These range from pregnancy (if you are a woman having sex with a man and don't want to be pregnant), potentially dangerous disease or mild irritation, heart attack, injury when the bed collapses, to hurt feelings – not even a phone call? – the chance of embarrassment or arrest if caught in a public place.

Let's start, as most of us did, with simple, unprotected sex between a man and a woman. What is the chance that any one of Kelvin's romantic encounters will end in a pregnancy? This too, for understandable reasons, is difficult to study in laboratory conditions. A New Zealand study in which participants were only allowed to have sex once a month suffered, unsurprisingly, from a high drop-out rate. Perhaps the closest has been a European study that recruited 782 young couples who did not use artificial contraception and who carefully recorded the date of every act (and there were a lot of them) until there had been 487 pregnancies.[2] The time of ovulation was estimated for each menstrual cycle.*

---

* The simplest way to estimate the chance of pregnancy was by considering only cycles in which there had been only one act of intercourse: the peak chance was observed at three days before ovulation, in which 8 out of 29 (29 per cent) 'coitions' resulted in pregnancy. But this only used around a third of cycles in which there had been any sex, and so a more sophisticated mathematical model was applied to the full data: they estimated the peak at two days before ovulation, with a chance of around 25 per cent, which is similar to previous estimates. The chances drop fairly steeply away from this peak, with an average of 5 per cent over the whole monthly cycle.

The bottom line is that a single act of intercourse between a young couple has on average a 1-in-20 chance of pregnancy – this assumes the opportunity presents itself on a random day, as these things sometimes do when you're young.

People who study how populations change are called demographers, and they use the term 'fecundability' for the probability of becoming pregnant during one menstrual cycle. This varies between couples, but the average is estimated to be between 15 and 30 per cent in high-income countries.

We can use the lower figure to illustrate the consequences for an average couple trying to have a child: there is an 85 per cent chance of finishing each month without conceiving, and if we assume each month is the same and independent, then there is a 0.85 x 0.85 x ... x 0.85 (12 times) chance of not getting pregnant in a year of trying, which is 14 per cent. So fecundability of 15 per cent means a 100 − 14 = 86 per cent chance of getting pregnant in a year: a figure of 90 per cent is often quoted as the proportion of young couples who will be expecting a child after a year without contraception, which corresponds to a fecundability figure of 18 per cent.

Fecundability can be estimated from large populations if there is no effective contraception. In Europe this means going into history: over 100,000 births taken from registers in France between 1670 and 1830 have been analysed to produce an estimate of average monthly fecundability of about 23 per cent.[3]

But suppose you don't actually want to get pregnant – how much is fecundability reduced by different types of contraception? This is generally expressed as the pregnancy rate following one year of use and strongly depends, of course, on how careful you are.

Contraceptive pills, intrauterine devices, implants and injections are quoted as 99 per cent effective, so that less than 1 in 100 users should be pregnant after a year, while male condoms are around 98 per cent effective if used correctly, and diaphragms and caps with spermicide are said to be 92–96 per cent effective, so between 4 and 8 women using them will be pregnant after one year.[4] Imperfect, or what's often referred to as 'typical', use of contraceptives – you threw up the pill in the gutter on

Figure 11: **Percentage of women pregnant after one year with selected forms of contraception**[5]

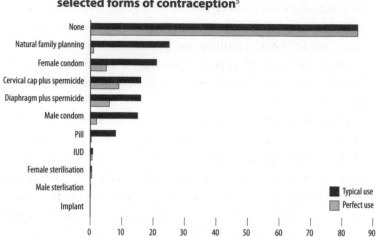

The cervical cap pregnancy rate shown here should be about doubled for women who have previously given birth. That is, the cap is only half as effective if you've already had a baby. The natural family planning 'perfect-use' rate varies from 1 per cent to 9 per cent pregnancies, depending on method. Failure rates for condoms are the figures without the use of spermicide. Other figures, from the National Institute for Clinical Excellence (NICE), indicate a failure rate for the contraceptive implant Implanon of less than 0.1 per 100 women over five years, better even than the numbers we've used here, and by some way the most effective form of contraception, according to the NICE data. Note that this doesn't address possible side-effects.

Friday night, forgot, it fell off, etc. – is, unsurprisingly, a lot less effective.

One way to make these comparisons is to imagine a woman considering various contraceptives. Assume an infinite sex life. How long before she typically becomes pregnant? In the case of female sterilisation, about 200 years. The failure rate is one pregnancy for every 200 woman-years. So if 200 women took it for one year, one would be expected to become pregnant. For contraceptive implants the failure rate is so low that an accurate figure is hard to calculate. By one estimate, it is about one pregnancy every 2,000 woman-years.

If they don't have access to contraception, then women can end up having a *lot* of children. The French data showed that women in the 1700s who married between 20 and 24 had on average 7 children each, as women still do in Niger and Uganda. *Total Fertility Rate* is the average number of children expected per woman if current fertility carried on through her life. It took the UK more than 200 years to reduce the number of births per woman from 5.4 in 1790 to 1.9 in 2010, while other countries have shown a similar decline in just a generation: Bangladesh took only 30 years to go from 6.4 in 1980 to 2.2 in 2010.* Some countries have a remarkably low fertility rate, particularly prosperous countries in South-East Asia (Singapore's is only 1.1) and countries in Eastern Europe – the Czech Republic's is 1.5, well below what is necessary to replace the population.

Pregnancy is not generally considered a great idea for young girls. In 1998 in England 41,000 girls aged between 15 and 17 conceived – that's 47 in every 1,000, or 1 in every 21. Imagine that at school. The UK government set a target to halve this rate by 2010, and by 2009 the headline figure had dropped to 38 per 1,000 girls, which is an important fall but a fraction of the target. There is wide variation in these rates around the country – from 15 per 1,000 in Windsor and Maidenhead to 69 in Manchester: that's 1 in every 15 girls aged between 15 and 17, pregnant every year. There is a strong correlation with low educational attainment, and deprived seaside towns traditionally have high rates, with 1 in every 17 pregnant in Great Yarmouth each year, and 1 in every 16 in Blackpool[7].

Nearly half of these teenage pregnancies end in abortions, but there are still many births. A 2001 report[8] put the UK highest in Europe, with 30 births per 1,000 women aged 15 to 19, with only the USA exceeding it among high-income OECD countries, at 52 births per 1,000. This is in staggering contrast to countries such as South Korea, Japan, Switzerland, the Netherlands and Sweden, which all have birth rates of fewer than 7 per 1,000 teenagers.

Of course, we could turn these dangerous odds around and make them the measure of hope for those who want to have a baby. Pregnancy isn't always well described as a risk – it can obviously also be a blessing.

Which you can't really say about disease and infection. HIV, syphilis, gonorrhoea, chlamydia, hepatitis and numerous other STIs – some potentially fatal, some unpleasant – are a danger of unprotected sex of various kinds, and some even with protection, whenever there is a chance your partner is infected.

---

* Hans Rosling's *Gapminder* depicts the trends vividly.[6]

Figure 12: **Historical number of diagnoses of syphilis in England and Wales (left-hand axis) and Scotland (right-hand axis) for men and women, GUM clinics**

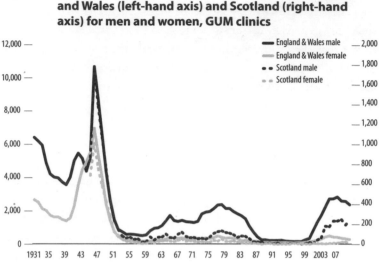

Risks are higher for young adults, men who have sex with men, and people who inject drugs, and black African and black Caribbean men and women continue to be disproportionately infected. The peak risk in women tends to be among those aged 19 or 20, and in men a few years later.

There has been an increase in the number of diagnoses for many STIs in the past ten years, partly because people's behaviour has changed, mostly because we are simply testing a lot more people through screening programmes and using more sensitive tests.

But the longer-run data are fascinating. Figures 12 and 13 reveal a rich history of behaviour linked to world events – wars, social shifts in attitude, AIDS and medical discovery – all expressed through sexual risk. Just look at 1946 in Figure 12, the chart on syphilis – and draw your own conclusions (noting that penicillin became widely available soon after).

The chance today that you'll catch something after one sexual encounter with an infected person is not a figure medical authorities like to throw around. It is extremely dependent on your partner's background, which disease they have and what you get up to.

Very roughly, the HIV risk for a woman with a man has been put at

Figure 13: **Historical number of diagnoses of gonorrhoea, England and Wales, GUM clinics**

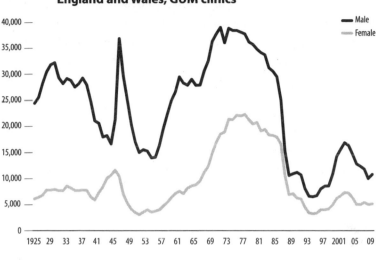

Source: Health Protection Agency, 2012

about 0.1 per cent (meaning that if she has sex with 100 infected men she will have a 1 in 10 chance of becoming infected herself); for a man with a woman it has been put at 0.05 per cent, or one infection per 2,000 'events'; and for a man with a man it goes up to 1.7 per cent depending whether you are 'insertive' or 'receptive'. But there are plenty of stories of once being enough.[9]

Again, these risks vary according to such factors as the strength of the virus in the infected person's bloodstream. For gonorrhoea, see Figure 13, the risk of infection has been reported as up to nearly 1 in 2 for heterosexual partners.[10]

And then we must remember that sex can be energetic, which itself carries a risk. It's been recently estimated that 1 in every 45 heart attacks is triggered by sexual activity.[11] Luminaries such as Nelson Rockefeller, Errol Flynn, President Felix Faure of France and at least two Popes are said to have succumbed this way. Solo sexual activity used to be associated with blindness and stunted growth, for which let's say there is a limited evidence base, but if it involves asphyxiation it is not recommended for the cautious. Numerous fatalities have been recorded, including David Carradine, Michael Hutchence and a British MP. A study recorded 117 deaths from just two states in Canada.[12]

Finally, sex has to be the moment we mention optimism. Kelvin isn't the sexually careful type (see the Fray Bentos metaphor/image). No surprises there. A variety of attitudes can complicate sex and risk, and he has plenty, not least a heroic lack of self-control (see eating Kath's flake).

The first of these attitudes is wish-fulfilment, behaving as if what you wish to be true, will be true, as in 'It'll all be all right'. This is funny to watch in young children – 'Can I drive the car, daddy? I won't crash.' – less so in adults, but still common. The puzzle is that we know we do it, and we still do it. As the Buddy DeSylva song 'Wishing Will Make it So' (from the film *Love Affair*) says: 'Wishing will make it so/Just keep on wishing and care will go.' This also goes by the glorious academic name of 'desired end-state' bias, in which you calculate that the risks are lower for no better reason than you would like them to be lower. 'You won't get pregnant. It'll be fine.' People are similarly reported to underestimate the chances of divorce or losing their job – because they don't want it to happen.

Call this positive thinking, but bear in mind the following examples, useful because they quantify the 'it'll be all right' tendency. First, the 'demonstrated, systematic tendency for project appraisers to be overly optimistic' during building projects. What could possibly go wrong with a building project? Plenty, as we all know. Project appraisers know this too. But still they tend to underestimate the risks – so much so that the UK Treasury forces them to make a formal adjustment for over-optimism.[13] So for capital spending on developing new equipment, even when you think you've thought of everything that could reasonably go wrong, add between 10 per cent and an astonishing 200 per cent for all that probably will. For the time it takes, add between 10 per cent and 54 per cent. Even for standard buildings, add up to 24 per cent to the projected bill; for non-standard civil engineering add up to 44 per cent to your best estimate of the cost.

A related problem is known as the planning fallacy, most famously demonstrated by 37 psychology students whose average estimate of the time it would take to finish their thesis was 33.9 days, or 48.6 days if everything went as badly as it possibly could. In the event, the average time was a whole week longer than that.

It's been suggested that we are 'hard-wired for hope',[14] since hope for a better future might encourage us to strive for it. If this is true, evolutionary fortune favours the optimistic. This is also a conjecture. For though it might be true that over-optimism helps Kelvin's genes spread far and wide, that's not what he's striving for. His is the wrong kind of optimism.

Another explanation for over-optimism is that you know yourself better than you know others. Asked to think about the risk that someone else will be accidentally knocked up and, lacking stats about average behaviour, we might try to picture a typical case, but really invent a crude caricature. Who is she, this slapper who typically gets pregnant? 'Well, she's drunk, she gets around, she just doesn't care, she's almost asking for it.' In other words, if we think about the risk of getting pregnant, we tend not to think of statistical averages, we tend once again to conjure up images of extreme cases. Then if we compare ourselves to these extremes, we might say: 'That's who it happens to, and I'm not like that, so I'll be all right.'

The moral of this? That given sufficient motivation – and sex is often a powerful one – risk is easily confused with morality, hope and even convenience, especially in the stories we tell ourselves.

9

DRUGS

PRUDENCE DRAINED HER COCKTAIL of butanol, iso amyl alcohol, hexanol, phenyl ethanol, tannin, benzyl alcohol, caffeine, geraniol, quercetin, 3-galloyl epicatchin and 3-galloyl epigallocatchin.

She felt the dark brown liquid flow around her mouth, sucked it down, swallowed, eagerly, gratefully. The chemical concoction seeped into her body, the liquid radiated warm, sedative comfort. She knew the effects; in her short life she had already come to depend on them. Her eyelids fell, she leaned back in her chair and let out a long, tranquillised sigh as her arm fell to the table with the empty cup.

'More tea, Pru?' said Norm.

A grey afternoon. And Kelvin had never felt greyer, or sicker. All colour gone, from him, from the parade of shops he walked past, traffic it hurt his head to look at as he crossed the road, mad visions of even his blood's redness leaching away.

Sensation gone too, except pain. Pain that began hours ago between his shoulder blades, spread through his ribs into a 10-tonne, coat-of-lead ache with every step. Moaning for relief through the metal taste in his mouth, the insomnia-lethargy, nausea, dehydration, irritability, anxiety, tearfulness and shot concentration, he needed another hit, now, to ease the pain, but he was all out. 'Now' was all he could think of. Another, now.

Kelvin spent his entire youth seeking instant gratification. At 18, he'd done enough mandy to ski in – bombing a half gram of powder at 5 a.m. at a festival one day and after that hooked, the mad looking-and-not-seeing and then seeing it all different and weird in the street lights. He loved it. He didn't stop until several years later, when his heart felt different, pumped different, the twitching hands and feet, and aches. So he quit pure, powdered MDMA, took something for the aches it bequeathed him, which became pains, took some more and now craved that instead.

He knew where to go and knew the routine. Be there, just before 6, with the money. So here he was, carrying his coat of pain, sweating. He ducked inside. The man he knew only as 'the man' was there, sure enough. Always the same man, same coat, same place, lurking in the back. They nodded. Kelvin walked over. The man had it. Kelvin handed him the cash.

Then the man said what he always said, the same taunt every time, knowing Kelvin would be back: 'Shall I put the repeat prescription and the receipt in the bag for you, sir?' he said.

'Please,' said Kelvin. And he left, cradling his Vicodin like a baby, fumbling at the foil inside the bag to pop some before he was even outside the chemists.

'And you took it why?' said Prudence on another occasion.

'Ah, well, you see ... balance of probabilities weighted by risk perception ...,' said Norm.

'Oh God,' she said.

It turned out that Kelvin had been goading Norm again, this time by saying Norm's idea of the wild side was two cups of tea.

'So the question was how to experience a risk sensation with modest objective risk,' Norm said. And after refining the options – how to maximise *frisson* per MicroMort, etc. – it came down to horse-riding ... or Ecstasy.

'Am I really hearing this?'

'Death is possible in either case, naturally, but the point is to stick to practical calculations on an objective basis.'

'Is it really?'

'Of course, procuring a tablet was going to be an issue, so I approached a man outside the Big Lurrve Club who was suspicious but helpful, picked up a Lucozade and *Horse and Rider* at Smith's and studied the form, so to speak.'

'And?'

'Well, do you see a horse? Serious adverse event every 350 exposures! Obviously, logically, I dosed up and blissed out,' said Norm.

'And did you have to play *Happy Hardcore Mix* quite so loud?'

PEOPLE HAVE TAKEN mood-altering substances ever since they had moods. Remains of opium poppy husks have been found in Neolithic settlements in Europe; natives of South America have long chewed coca leaves as a mild stimulant or to suppress pain; almost everything edible has been fermented to make alcohol.

There are many drugs, and about every drug's risks and benefits there are many opinions. In this chapter we'll sample that opinion, one of the widest, wildest and most diverse spreads of opinion about any risk,* to see how diversity squares with numbers.

The arguments have been around long enough. Opium in the 18th and 19th centuries was variously described as an inspiration, a medicine and the devil. Some writers romanticised it. Samuel Taylor Coleridge took laudanum (from Latin *laudare*, 'to praise'), a tincture of 10 per cent opium in alcohol, before writing the poem *Kubla Khan* (1797), 'composed ... in a sort of Reverie brought on by two grains of Opium taken to check a dysentery' – until famously interrupted by a person from Porlock.

By contrast, John Jasper in *The Mystery of Edwin Drood* (1870), by Charles Dickens, awakes in a squalid opium den alongside a drooling Lascar, a haggard woman and a Chinaman convulsed by gods or devils. Jasper is a choirmaster, but in Dickens's hands opium betrays his degeneracy.[2] In Oscar Wilde's *The Picture of Dorian Gray* (1890) 'There were opium dens where one could buy oblivion, dens of horror where the

---

* See, for example, David Nutt[1] for a vivid description of different views on illegal and legal drugs.

memory of old sins could be destroyed by the madness of sins that were new.'

In politics and business too fashions changed, and not just towards prohibition or censure.* Drug production went from cottage industry to industry proper, beginning in about 1827 with Heinrich Merck's commercialisation of morphine, an extract of opium, laying the foundation of the Merck pharmaceutical company. Britain, as now, had a trade imbalance with China, and whereas today Britain exports malt whisky to the Chinese, in the 19th century the East India Company fought two wars for the right to ship them opium.

Heroin, also known as diamorphine, was first derived from morphine in St Mary's Medical School in London in 1874, but was rediscovered by the Bayer pharmaceutical company in 1897 and marketed as a non-addictive painkiller and cough medicine. Meanwhile cocaine was extracted from coca leaves. 'Cocaine toothache drops – Instantaneous cure!' was one product for children. Sigmund Freud was one of many cocaine enthusiasts, as was Sherlock Holmes, to Dr Watson's disgust:

> 'Which is it to-day?' I asked, 'morphine or cocaine?' He raised his eyes languidly from the old black-letter volume which he had opened. 'It is cocaine,' he said, 'a seven-per-cent solution. Would you care to try it?' 'No, indeed,' I answered, brusquely.[3]

Today a modern Holmes equivalent, Dr Gregory House in the medical TV drama *House*, can use and abuse Vicodin, a painkiller and opiate, just like Kelvin, even as politicians speak of a war on drugs. Romance and heroism still sit side-by-side with the end of civilisation, amid the stars and in the gutter.

Compare the following edited extract, from the oral testimony at a meeting of Narcotics Anonymous of a man who lost his daughter and

---

* A recent example of drug liberalisation is the Licensing Act of 2003, which relaxed the rules on the opening hours of licensed premises, even allowing continuous opening for some premises, in England and Wales in 2005 (similar changes were introduced in Scotland in 2006).

brother to drug abuse, in which drugs are the mark of a life almost born to ruin –

> Because me dad was always kicking fuck out of me mother and there'd be blood on me and blood on me brother an that and I'd be hiding me brother under the bed. And so I got out when I could. And what me dad done is he said I'd never be anything and he put all his hang-ups and all his issues and all his negative perceptions on me. And the way I coped with that is I became very paranoid and very angry with anyone who slighted me and I learnt how to be violent and how to get amongst people and hide really well. I started using when I was eight. Robbing Diconal off the prostitutes ... And so when I was 20 I finished up doing 13 years in high security and I didn't know who I was so anyone who had any identity I mixed with them. I was hangin' out with the Palestinians and the Moslems and I grew a beard. I was just taking on anyone's grievance as my own. And what me mum said about that she said, 'Better bring him some gear, straighten him out.' (laughter)

– with this, from Hunter S. Thompson, drug glutton and author of *Fear and Loathing in Las Vegas*, who had a journalistic style once described as extreme participatory fieldwork –

> I like to just gobble the stuff right out in the street and see what happens, take my chances, just stomp on my own accelerator. It's like getting on a racing bike and all of a sudden you're doing 120 miles per hour into a curve that has sand all over it and you think 'Holy Jesus, here we go,' and you lay it over till the pegs hit the street and metal starts to spark. If you're good enough, you can pull it out, but sometimes you end up in the emergency room with some bastard in a white suit sewing your scalp back on.[4]

The point again is the variety of view – from deadly to life-affirming. Ideas of what it is that's at risk have also shifted: it's not just health at stake now but relationships, employment, crime, rainforests cut down to farm coca – far away from the drug-taker.

To many, drug abuse is used needles, squalor and everything the film *Trainspotting* is about. To others, it comes with a corporate logo. Some addictive drugs – painkillers like Codeine, for example – are available over the counter; others – alcohol at the pub – may be woven into a way of life. Cathryn Kemp 'used to think a drug addict was someone who lived on the far edges of society. Wild-eyed, shaven-headed and living in a filthy squat,' until she become one.[5] What began as a legitimate need for pain relief and a painkiller called Fentanyl became a comfort as she consumed it the way she'd take a glass of wine to unwind at the end of the day, then moved to dependency, then to about ten times the maximum recommended dose, then to thinking only of the next hit, living only for the drugs, raging at her family, finally accepting that she was an addict and a brutal rehab, all via nice packaging and clean suppliers at local chemists, much like Kelvin.

The Times reported in 2012 what it called the 'Scandal of 1m caught in tranquilliser addiction trap', about 'the failure of the NHS to provide meaningful support to victims of benzodiazepine addiction, even though they significantly outnumber those hooked on illegal drugs'. Unlike with heroin, there are no hard-hitting public information campaigns about how painkillers screw you up.

Are Prudence, Norm and Kelvin all users? They would disagree, as will readers. They would also differ over the severity of the risks. Which is worse: booze, Kelvin's painkillers or cocaine? Does Pru's stimulant of choice (tea, but not Fair-Trade tea – so some will say that her tea addiction is socially harmful) belong on the same page as Norm's experiment with E, or is Ecstasy little more dangerous than a cuppa for many self-described recreational users who successfully hold down a job and relationship?

Knowledge changes, beliefs vary, attitudes are all over the shop and numbers are only part of it. People's sense of harm is and always was plainly bound up with personal experience, anecdote, moral values and personal preference – what's your poison? – as well as social norms, norms among friends, among their income, ethnic and age groups, or in society as a whole, norms that are sometimes freely chosen and sometimes not: whether, for example, they turn into social pressures to conform, with a drink, or a tab.

Often the illicit drug-user's first line of defence is to say that attitudes to alcohol prove there's no consistent scale of harm, only shades of hypocrisy. In the decades since governments began to talk more about the dangers of drug use, two of the most potentially dangerous and addictive drugs – tobacco and alcohol – remained legal, even as drinking in the UK rose sharply in the 1980s and early 1990s. As a result, it is said: 'A whole generation learned to ridicule and ignore all governmental advice on the subject.'[6] Others talk – not always in ironic tones – about 'respectable drug addiction' – middle-class use of the opiate-based Codeine to help with sleeplessness, for example. What could be less sinister than wanting a good night's sleep? But Codeine dependency can be ugly and dangerous too. Rules and attitudes said to be linked in large part to risk and harm have a fickle relationship with it.

Even so, and despite being keen on data, we would argue that little of this mess is intrinsically wrong or irrational, although it may not always be honest or self-aware. It is simply that danger is only one part of a hideously complicated conversation. How do we describe the risks of drug use when they are part and parcel of what we value in life? Because this is a problem of weighing both personal harm and social harm, harm to a way of life, where sometimes that way of life is threatened by drugs – by the crime they bring – but sometimes it may depend on them, like a few pints down the pub at Sunday lunch, half an E at a club or a cigar at the races. Add to that a fierce disagreement about how best to minimise the harm, by prohibition, decriminalisation or a war on drugs, and you have a recipe for muddle.

Questions of value give philosophers headaches. The point of saying all this is that risk can also fall – and should – into the same category of philosophy, one of competing values and traditions at least as important as the power of the data. So heated is the argument about illicit drug use that it is almost surprising anyone should think data could settle it.

Perhaps the acid test (no pun intended) of the relevance of the data to this whirl or beliefs and emotions is how you react to Norm. What did you think when he tried to skip over all the values business with a purely data-driven approach to excitement, illegal or not? Is he the only sensible person on the planet? Prudence doesn't think so. Or is his kind of logic, in this instance, insane?

The anthropologist Mary Douglas argued that the assertion of risk – 'don't do that, it's dangerous' – is often sly social control. If behaviour offends us, for whatever reason, we warn that bad things will happen and we call the behaviour 'risky'. In her fieldwork she found tribal women who'd been told that if they were unfaithful they ran a greater risk of miscarriage, a risk presented to them as a fact of nature a bit like the need to take care with a sharp knife. The supposed biological risks of the behaviour were clearly a fiction; the true purpose, the social purpose, of warning about the risk of infidelity was evident: to control the women's behaviour. Early in her career Douglas thought this the kind of thing only primitives do, since superseded by science. Later she argued that we all do it, in all cultures and all ages.

But wherever one person seeks control, another kicks back. Hunter S. Thompson couldn't consume enough illicit drugs, more determined to stuff them down because others wanted to stop him. His accounts of his drug binges read like a knock-down fight with majority opinion. He describes his 'twisted' – a good word – trip to Vegas as fulfilment of the American Dream. Some dream.

Thompson was not your everyday risk-taker. But was he unusual when he suggested that the best way to resist control is to assert control of your own, even unto being out of it? Transgression is partly what risk-taking is about, at least for some. Stepping over a boundary can be a thrill in itself.

So in front of any statistic about the quantified level of danger we need a flexible sign, positive or negative for different readers, depending on what they approve of, while also recognising that the value they put on any risk isn't only up to them. Their behaviour affects others. So calculating the dangers of drugs is both a personal judgement and a blazing social row replete with wider values and preferences at least as much as it is a statistical, evidence-based exercise.

Mind you, once sociology, anthropology, psychology and the rest have said all this, you sure feel the need for some numbers – if not to make policy, then at least to help make up your own mind. What are they?

Concern about the addictive properties of opiates and cocaine in the early 20th century led to criminalisation for misuse. Criminalisation makes it hard to estimate the level of harm from drugs, because you first need to know who uses them. If admitting this makes you a criminal, you might keep quiet.

So the British Crime Survey (BCS) guarantees anonymity when it asks about the use of illegal drugs.[7] Responses are scaled up to the adult (ages 16–59) population of England and Wales, of whom around 1 in 3 are estimated to have used illegal drugs in their lifetime, around 9 per cent in the past year.

About 3 per cent, a million people, had used a Class-A drug, such as heroin, morphine, metamphetamine, cocaine or crack cocaine, Ecstasy (MDMA) or LSD. Among young people aged 16 to 24, around 20 per cent had used illicit drugs in the last year: 17 per cent used cannabis, 4.4 per cent powder cocaine, and 3.8 per cent Ecstasy – the heroin figure of 0.1 per cent (1 in 1,000) is probably too low, owing to less than full responses from this community. Compared with 15 years previously, in 1996, there had been a fall in overall drug use, with a substantial decline in the figure for cannabis but an increase in cocaine and methadone. Men are about twice as likely to be users as women, and an unsurprising relationship was found between night-club and pub visits and illicit drug use.

What's so bad about these drugs? They can certainly be dangerous in the wrong hands: a Manchester general practitioner, Harold Shipman, injected over 200 of his patients with lethal doses of diamorphine (heroin) in a murderous career before he was finally caught in 1998 after a clumsy attempt to forge the will of one of his victims. And there have been numerous famous deaths from drug misuse, whether deliberate or not – from Janis Joplin to the Singing Nun.

But working out exactly how many people die this way is tricky: in general, if drugs are mentioned on the death certificate, it means the death is counted in the official data as drug-related, even if the drug was not the sole cause.[8] Altogether there were 1,784 deaths in England and Wales in 2010 from misuse of illegal drugs (i.e., not including alcohol), down slightly from preceding years but double the figure in 1993.

The peak decade for men is the thirties, with 544 deaths. That is around 150 MicroMorts per year, 3 a week, averaged over everyone in that age group, from dopeheads to vicars. Almost exactly half the total deaths (791) were due to heroin or morphine. Cocaine was associated with 144 deaths, amphetamine 56, while those involving Ecstasy fell to only 8 after averaging around 50 a year from 2001 to 2008.

If we use the BCS estimates of the number of users, we get a rough idea of the annual risk, in MicroMorts, for users of different drugs. Averaging out from 2003 to 2007,[9] cocaine and crack cocaine were involved in 169 deaths per year, and so an estimated 793,000 users were each exposed to an average of 213 MicroMorts a year, around 4 a week.

Ecstasy's 541,000 users experienced around 91 MicroMorts a year each, or around 2 a week. Since the 2003 market for Ecstasy has been estimated as 4.6 tonnes,[10] this corresponds to around 14 million tablets, or an average of around 26 per user. This translates to roughly 3.5 Micro-Morts per tablet.

Cannabis seldom leads directly to death, but its estimated 2.8 million users suffered an average of 16 associated deaths per year, which is 6 MicroMorts a year.

This level of harm is as nothing compared with the average of 766 heroin-related deaths a year, which works out at 19,700 MicroMorts per year – 54 a day – like a 350-mile motor-bike ride every day, or the exposure *every day* to seven times the risk from a year's worth of cannabis, although this very high figure depends on the BCS underestimate of users, and so is an overestimate of the risk.

But there are many other harms apart from death: for example, it's been estimated that smokers of cannabis are about 2.6 times more likely to have a psychotic-like experience than non-smokers.[11] Heroin injectors may contract HIV or hepatitis from non-sterile needles. They may have abscesses, suffer poisoning from contaminants, apart from the risks of dependency and withdrawal, not forgetting the standard effect of opiates on bowel movement.

John Mortimer (to his father): Did you ever smoke opium?
His father: Certainly not! Gives you constipation. Ever see
a portrait of that rogue Coleridge? Green around the gills and
a stranger to the lavatory.

[*A Voyage Round my Father*]

Potential effects ripple out from bad guts to violence to felled rain-forests and beyond. This is another way of stating the problem about how to make comparisons between harms that matter differently to different people. Is it possible? It's been tried, let's put it that way, including an attempt to put a single, summary nastiness number on each drug. A recent study that looked at harms of various drugs, including the legal ones alcohol and tobacco, took in such effects as mortality, damage to physical and mental health, dependency and loss of resources and relationships, as well as harms to society, such as injury to others, crime, environmental damage, family adversities (such as harm to family relationships), international damage (like those rainforests), economic cost and effects on the community.[12]. Each drug was scored on each dimension, the different harms weighted according to their judged importance and a total harm score calculated. As with any composite indicator, what goes into the mix and the weight accorded any single factor is arguable.

The resulting ranking put alcohol at the top, with 72, then heroin and crack cocaine, at 55 and 54; tobacco was sixth, at 26, and Ecstasy almost at the bottom of the list with 9, despite being a Class-A drug in the UK. This ranking was controversial, to say the least.

More controversial even than comparing illegal and legal drugs is to compare illegal drugs with wholesome activities. Professor David Nutt, then head of the Advisory Council for the Misuse of Drugs (ACMD), wrote a paper comparing Ecstasy with 'equasy', the addiction to horse-riding, claiming that both were voluntary leisure activities for young people of comparable danger.[13] He did not remain chairman of the ACMD for long. Not because his numbers were absurd. They weren't. But risk, to repeat ourselves, is not simply, or even mainly about the threat of harm. His political bosses seem to have decided that comparing horse-riding to drugs wasn't good politics.

Professor Nutt said the dangers of equasy were revealed to him by the clinical referral of a woman in her early thirties who had suffered permanent brain damage as a result of equasy, leading to severe personality change, anxiety, irritability and impulsive behaviour, bringing bad relationships and an unwanted pregnancy. She lost the ability to experience pleasure and was unlikely ever to work again.

He wrote:

> What is equasy? It is an addiction that produces the release of adrenaline and endorphins and which is used by many millions of people in the UK including children and young people. The harmful consequences are well established – about 10 people a year die of it and many more suffer permanent neurological damage as had my patient. It has been estimated that there is a serious adverse event every 350 exposures and these are unpredictable, though more likely in experienced users who take more risks. It is also associated with over 100 road traffic accidents per year ... Dependence, as defined by the need to continue to use, has been accepted by the courts in divorce settlements. Based on these harms, it seems likely that the ACMD would recommend control under the Misuse of Drugs Act perhaps as a class A drug given it appears more harmful than ecstasy.[14]

He concluded that a more rational assessment of relative drug harms was possible, but the drug debate took place without reference to other causes of harm in society, which gave drugs a different, more worrying, status.

Is that status deserved? If we think so, we need to say why, without saying that it is because drugs are more harmful, since harm alone doesn't bear the weight of opinion. Quantifiable risk is not the answer.

The final warning comes from Dr Watson as he witnesses his friend Sherlock Holmes indulging in a (then perfectly legal) drug, in *The Sign of Four*:

'But consider!' I said, earnestly. 'Count the cost! Your brain may, as you say, be roused and excited, but it is a pathological and morbid process, which involves increased tissue-change and may at last leave a permanent weakness. You know, too, what a black reaction comes upon you. Surely the game is hardly worth the candle. Why should you, for a mere passing pleasure, risk the loss of those great powers with which you have been endowed?'

10
BIG RISKS

KELVIN PEELED AWAY DOWN a steep path, marched to the door, scuffed to a stop, and knocked. Norm scampered after. He had a low opinion of people who disagreed with him. Dodgy motives the lot. Climate change was an excuse for wacko eco-zealots to tell you what to do. Activists were control freaks, and Greenpeace gave him Tourette's. As for all that stuff about polar bears ... blackmail with a furry animal.

'And whales should be used up,' Kelvin said. 'Why not? We need soap.'

A woman in a wide skirt, hazy through frosted glass, swayed down the hall, the latch clicked, the door half-opened, and through the gap came a cautious smile with a grey bob, somewhere in her late sixties.

He and Norm were in one of those bad first jobs that people do who are desperate for a first job, trudging door to door, hating suburbia, failing to drum up business and putting the world to rights instead.

Norm had dared to suggest that majority opinion accepted the case for anthropogenic climate change and Kelvin should consider the balance of probabilities, but Kelvin said that only showed people were f... wits, and to prove it...

'Good afternoon, my name is Mr Poe,' said Kelvin, offering his hand through the gap. 'And this is my associate, Mr Edgar.'

'Oh yes?'

'From London Zoo.'

'Oh?'

'You'll have seen the programme – on the television last night.'

'Which one was that?'

'About the zoo. The penguins.'

'I watch the BBC. For *EastEnders*.'

'Of course. But you'll know, you'll have heard about the penguins.'

'Well, I don't know. What about the penguins?'

'Being closed.'

'Closed?'

'That's why we're here. Closed down. Climate. Habitat. It's an absolute urban heat sink in Regent's Park, you know.'

'Oh. What will they do?'

'Quite. That's why we hope you can help. That's what the appeal is about.'

'An appeal, I see.'

'You're most understanding,' said Kelvin, who advanced like a vicar to the needy and took her hands. 'And you do have a nice bit of garden at the back there, high up, flood-free.'

'Oh yes,' she said, letting go of his hand.

'There you are then. Just right isn't it, the garden?'

'Well, I like it.'

'Excellent. I'll put you down for two.'

'Two?'

'In the morning.'

'I'm sorry, I'm not buying ...'

'No, no, they're free, completely free, though in the circumstances you're the generous one.'

'Well I, I don't ...'

'No need to worry, full instructions with every penguin. About eight o'clock all right?'

'Well, no ...'

'Two penguins to number 17, Mr Edgar.'

'What?'

'Penguins, Mr Edgar.'

'There's been a misunderstanding. I'm sorry ...'

'Plenty of shade,' said Kelvin turning away. 'And they like muesli.'

'But I eat corn flakes.'

'Cornflakes will do. But not too much or they get fat. Come, Mr Edgar, sixty-four to go.'

With a nod: 'You're so kind. The penguins will be most happy here.'

He marched.

'No, you can't ...'

Up the path, Norm trotting after.

'Really. This can't ... excuse me ...'

Brisk, 'You!' ... and 'You!' ... gone.

At 08.39 GMT the following morning Richard Kowalski of the Catalina Sky Survey made observations using the Mount Lemmon 1.5-metre aperture telescope near Tucson, Arizona.

He was puzzled. The potential near-Earth object SO43 didn't seem to be quite where he expected. An error, probably. He emailed the Minor Planet Center in Cambridge, Massachusetts, to let them know they might like to check the data.

WHAT DOES SUCKERING an old lady about penguins have to do with big risks like climate change?* Plenty. But to see it, we have to step outside our own beliefs about these risks. So, just for a moment, forget whatever you think is the truth about climate change and accept that, as risks go, opinion touches the extremes – from nothing much to worry about, to the end of life as we know it. Is global warming a hoax, as Kelvin thinks? Or are we going to fry? People disagree.†

---

* By big risks we mean those that affect lots of people, such as the climate, new diseases, natural disasters and so on.

† 'Sharp decline in public's belief in climate threat, British poll reveals,' *The Guardian*.[1] ''Thirty one per cent said climate change was "definitely" happening, while 29% said "it's looking like it could be a reality", and another 31% said the problem was exaggerated, a category which rose by 50% compared to a year ago. Only 6% said climate change was not happening at all, and 3% said they did not know.' About a year later, the paper reported that 'Public belief in climate change weathers storm, poll shows', with 14 per cent 'saying global warming poses no threat'.

One side say the great majority of scientists who study climate change think the planet is genuinely warming, that people are responsible for it, and it's serious. The sceptics on the other side are wrong, conspiracy theorists, nutters in the pay of ... etc.

Sceptics like Kelvin appeal to science too, and return the abuse. The great majority of proper, competent, honest scientists think human-made climate change is hot air, they say. The rest are conspiracy theorists, nutters, in the pay of ... etc.

They say this about even the most basic question of the lot: is the world getting hotter? Forget whether people are responsible for it, or what to do about it, just that simple, measurable detail: is the world getting hotter? Still people disagree. Still they think the science – the real science – is on their side.

Which is not to argue that climate change is all a matter of opinion. But it does show how opinion about danger can work wonders with much the same evidence. So what drives the views people take about what ought to be – in an argument where everyone says evidence is paramount – basic scientific facts?

We could say the answer is trivial and obvious: that everyone is biased. People see what they want to see, scientific facts included. But the beginning of a more interesting answer is that the facts as we see them about big risks like climate often depend less on the facts than on who we are. Kelvin's politics, love of freedom, irritation with government, tendency to risk-taking and impulsiveness, even his loathing of the conformist wasteland of suburbia, are not coincidentally linked to his beliefs about the facts of climate change. They directly contribute to it. They also feed his scorn for people who can be taken in by the 'lies' of the other side. That is, our views of the facts about big risks are often prompted by our politics and behaviour, even as we insist that the rock on which we build our beliefs is scientific and objective, not the least bit personal, even as we swear we believe what we believe because it's true, why else? At least, that's the story we tell ourselves. It's everyone else who is swayed by personal and political baggage, and the story we tell of them is about their stupidity and corruption.

But it is not just the evidence offered by the other side that we dislike or disagree with; it's what those facts seem to tell us about the other side's whole political make-up. That's why Kelvin feels free to be rude to an old lady: if she's willing to believe all this penguin shite, she's probably a socialist.

We have already said, in the chapter on drugs, that values can drive people's sense of risk. Climate is in some ways similar, but a tougher case to prove, since few people are willing to admit that their views on the risks of climate change, or even the basic fact about whether the world is getting hotter, arise from their political make-up. But that is the argument we will consider in the chapter.

A quick word of caution: if you think climate change is proof that cherry-picking scientific evidence is typical only of the political right, hold fire. Those who study these habits say all sides have them, depending on the risk at issue.

A common belief among scientists is that people who seem to them to hold views inconsistent with the evidence simply need enlightening with the facts. But, if the argument in this chapter is right, their attempt will fail and probably backfire.

Here's how the process works. Because the climate is a huge, complicated system, it is impossible to be certain what it's up to. Persuaded, yes; certain, no. And the smallest doubt creates room for disagreement. This is where values creep in. To see how, let's take a quick digression via engineering at the micro scale known as nanotechnology – 'grey goo' to its critics. What do you know about nanotechnology? If the answer is anything like as much as MB and DS, not much: a bit of rumour, a bit of mystique, a lot of syllables.

Next question: how do you *feel* about nanotechnology? Again, if you're like most people studied on this question, despite an absence of facts you will probably have a view: 'excited' maybe, or 'anxious'. You may even have a few speculative reasons for the way you feel. Go ahead, be our guest, sound off, ignorance is no bar. Most of us do to some extent. We form our initial opinions about danger with the haziest grasp of the known facts.

Dan Kahan, at Yale University, leads what he calls the Cultural Cognition Project, studying how people form opinions about risk [2] Kahan tells the story of the reaction in 2006 of the city authorities in Berkeley, California, to a proposal by the university for a nanotechnology research facility:

> Having never heard of nanotechnology before ... the city's
> hazardous waste director immediately commenced an inquiry
> ... 'We sent them a bunch of questions starting with: "What
> the heck is a nanoparticle?"' Regulators were quickly able to
> learn that, but not much more: 'The human health impacts
> of nanoparticles', the city's Environmental Advisory Commission reported, 'are very complex and are only beginning to be
> understood.' Nevertheless, citing concerns that nanoparticles
> might 'penetrate skin and lung tissue' and possibly 'block or
> interfere with essential reactions' inside human cells, officials
> concluded that a precautionary stance was in order.[3]

Knowing next to nothing, the city regulator still formed a view about the risks. He could decide, as he did – as can any of us – that things are risky until there's a hint otherwise. Or he could have decided that things are safe until there's reasonable evidence otherwise. So what made him jump one way rather than the other? Not the evidence, says Kahan, there wasn't much, either to suggest that nanoparticles were dangerous, or that they were safe. The deciding factor was his cultural disposition.

Funnily enough, whatever people's first instinct about risks like this, more information only seems to confirm it. Kahan and his associates went on to question another 1,800 Americans about nanotechnology, discovering that it was their visceral, emotional response, pro or con, that determined how risky they thought it and that these views generally only hardened the more they learned, suggesting that when we claim to take a provisional view it is often a pretence to others and maybe to ourselves that we're open-minded when our opinion is already a closing door, swung not by facts but a set of prior cultural attitudes.

Kahan's work suggests that people tend to assimilate new knowledge 'in a manner that confirms their emotional and cultural predispositions'. In other words, they filter the facts to suit their beliefs and instincts from the first. Belief doesn't simply follow fact, belief decides what the facts are. So, is nanotechnology dangerous? Ask instead: 'How do you feel?'

Sometimes people say that they just don't know what to think, it's all too complicated. So we turn to the experts. But still we decide – who'd have thought it? – that most experts give answers consistent with the way of life and politics we happen to like, and we do this by picking and choosing who qualifies as an expert. If you're a hippie with a wild beard and I'm a suit with a short hair-cut, I'm less likely to rate your qualifications, whatever they are. We judge the experts' expertise according to whether they seem to be people like us, people who share our cultural outlook. Again, Kahan's experiments seem to confirm this. He shows people pictures of made-up experts – including one with a wild beard and another with a suit and short hair-cut – all with impeccable academic credentials, to see what it is that people recognise as expertise. They choose carefully. Social psychologists call this biased assimilation. Being the serious sort of person who reads books like this, you're above all that, of course. Except that Kahan finds that, the more scientifically literate people are, the more they do it.

He argues that we can all be placed on a spectrum of attitudes and beliefs,[4]* and once he knows how you feel about nanotechnology, he has

---

* Kahan's work here draws heavily on the cultural theory of risk. One of the most influential accounts of this is in *Risk and Culture*, by Mary Douglas and Aaron Wildavsy, which describes arguments in the US about nuclear power and air pollution in terms of competing ways of life: on one side egalitarian collectivists who use fear of disaster to argue against the kind of free enterprise that they also believe brings inequality; on the other side hierarchical individualists who want to defend free enterprise from public interference.[5] Douglas argued that the various psychological ways of viewing risks that we look at elsewhere in this book are just tools people use to assert their cultural outlook, the means to an end. She was more interested in the ends. See p. 101 for Mary Douglas's ideas about risk as social control in Chapter 9, on drugs.

See also *The Righteous Mind*, by Jonathan Haidt,[6] who argues that our views about subjects such as climate change – along with most other political issues – come from a small set of tribal, moral outlooks. These have much in common with Dan Kahan's use of cultural theory. 'Morality binds and blinds', says Haidt, a good idea how you will feel about the risks of climate change, nuclear power, gun control and so on. What's more, he says, you will believe, wherever you stand, that the true scientific consensus – not the nutters and conspiracy theorists – is with you. This is what Kahan means when he talks of cultural cognition.

One way of classifying people is to put them on a line that runs from 'individualist' to 'communitarian'. Individualists tend to dismiss claims of environmental risk, 'because acceptance of such claims implies the need to regulate markets, commerce, and other outlets for individual strivings,' says Kahan, just as Kelvin doesn't believe the Green movement because he thinks that every time it sees a crisis it starts telling people what not to do.

Communitarians, on the other hand, resent commerce and industry as 'forms of noxious self-seeking productive of unjust disparity, and thus readily accept that such activities are dangerous and worthy of regulation', says Kahan.

None of this is unfailingly accurate, of course; there are plenty of exceptions. We are not arguing that climate-change sceptics have a monopoly on rudeness. But part of Kelvin's reason for rejecting climate change is that he doesn't like the political colour of the solutions, which tend to mean more government, more regulation, more criticism of private enterprise. And if that means more control, it can't be true. Those on the other side, Kahan would say, believe what they do about climate change not necessarily because they're more scientific, but also because this gives them a chance to kick private enterprise.

Big risks – those that might endanger hundreds, or thousands, or millions – are prone to cultural cognition, partly because the evidence is bound to be uncertain to some extent, as there's no way of getting more data by re-running experiments on the whole planet with and without an industrial revolution.

And partly because the crunch may be years away, affecting our children more than us – and people will have different views about how

---

meaning that morality, rather than scientific evidence, is what keeps your side together and what makes the other side seem stupid.

much the future matters. So climate change, like every other risk, is affected by the probability/consequence distinction, where some will care more and others less about the potential costs to future generations. In fact, we can express our feelings about the future in a simple number, what's known as the 'discount rate'. If something that will happen in 50 years' time is just as important to us now as something that happens now, then our discount rate is 0 per cent. We don't discount the future.

But this doesn't describe most of us. If you're offered £5 now or next month, most people think the £5 in the future is worth less – because it's in the future. They discount it. When deciding on health policies, the UK National Institute of Health and Clinical Excellence (NICE) uses a default discount rate of 3.5 per cent, which means a year of life in 20 years' time is worth only around half of a year of life now. Hospitals deciding whether to use their limited cash to treat an older person or a younger one could say that the young person has a bigger future, and this counts for more. But if that future is heavily discounted, this works in favour of older people because doctors are now required to put a higher value on their current year. Not having much of a future simply by virtue of being old is then less of a disadvantage in the scrap for resources. With a 0 per cent discount rate – which implies future years are as valuable as current years (and the future is not discounted) – all the NHS money would be spent on younger people, who have more years left in which to benefit.

When economists look at climate change, they usually use a lower discount rate, such as 0.5 per cent – to reflect people's concerns for the long-term future. Similarly, when deciding where to dump nuclear waste, if we had a high discount rate, we might just shove it in a bucket in a cave where it would probably be OK during our lifetime. But which discount rate is the right one can't be settled by science or statistics.

The UK National Risk Register[7] gets around this by only being concerned with horrible things that might happen over the next five years. People sit around mulling on possible causes of death and disaster and cheerily try and assess how bad they might be and how likely they are, and put them into a table like Figure 14 (overleaf). The scale of harm goes from 1 to 5 on the vertical axis. This is a human judgement, not a statistical fact, and so is the likelihood that many of these events will come to pass. In fact, the best that can be said in defence of these numbers is that they are better than nothing for governments that have no choice but to set priorities and make policies. They are intended to be fair working assumptions, not predictions.

Figure 14: **UK National Risk Register**

How serious	1 in 20,000 to 1 in 2000	1 in 2000 to 1 in 200	1 in 200 to 1 in 20	1 in 20 to 1 in 2	Above 1 in 2
5				Pandemic influenza	
4			Coastal flooding / Effusive volcanic eruption		
3	Major industrial accidents	Major transport accidents	Other infectious diseases / Inland flooding	Severe space weather / Low temperature and heavy snow / Heatwaves	
2			Zoonotic animal diseases / Drought	Explosive volcanic eruption / Storms and gales / Public disorder	
1			Non-zoonotic animal diseases	Industrial action	
			How likely		

These big hazards are hard to quantify with anything like the accuracy of the risk of heart disease, but they are risks that alarm governments. And again, it's clear that how serious you think they are may depend on your cultural outlook. Are you horrified by the threat of industrial action – militants bringing the country to its knees – or do you see no risk there at all, just ordinary people defending their living, a threat to no one if they're treated fairly? This is a repeated and important point in this book and in studies of risk generally: that some arguments about risk are not really about risk.

The UK National Risk Register only puts these threats into broad categories. To say, as it does, that the risk of an 'explosive volcanic eruption'
is between 1 in 200 and 1 in 20 means anything between a 0.5 per cent
and a 5 per cent chance over the next five years. But there isn't much to
base more precise figures on: the volcano in question in Iceland has gone
off only twice in the last thousand years, so if we say that very roughly it
goes off on average every 500 years, that's a 1 per cent chance it will go off
in the next five years. Mind you, it ranks highly on 'impact': the last eruption, in 1783, killed 20 per cent of the population of Iceland, wiped out
their agriculture, sent clouds of sulphur dioxide across Europe, which fell
as sulphuric acid and caused years of bad harvests, and so contributed to
the French Revolution in 1789. One to watch out for.

When one of these crises occurs, or comes close, somebody has to
be wheeled out to tell the public about it. There are manuals to teach
you how to do this,[8] full of commandments such as: listen to people's
concerns, build on trust, express empathy, act fast, repeat messages, tell
people what to do, acknowledge uncertainty, don't just reassure, commit to learning.

Not much of this advice seems to have been followed when there was
an outbreak of severe food poisoning in northern Germany in early May
2011, including cases of lethal hemolytic-uremic syndrome (HUS).[9] A
local laboratory tested some organic Spanish produce, found E. coli, and
on 26 May announced they had found the source. Result: a widespread
boycott of Spanish vegetables despite film of the Spanish Minister of
Agriculture desperately munching on organic cucumbers to show how
safe they were. The problem was that, although the laboratory genuinely
identified some E. coli, it was not the type that was causing the casualties, of whom 50 died. It was only once the Spanish industry had been
well and truly messed up that the source was finally tracked down, in late
June, to a shipment of Egyptian fenugreek seeds. And who remembers
that?

An even more chilling case-study in how *not* to communicate risk
was the earthquake in L'Aquila, Italy, in 2009. After a series of small
shocks and a local amateur with home-made equipment had predicted
a major earthquake, there was a crucial meeting of experts on 31 March

2009, intended by the Civil Protection Agency to reassure the public. The meeting concluded that 'there is no reason to say that a sequence of small magnitude events can be considered a sure precursor of a strong event', but at a press conference afterwards an official, Bernardo De Bernardinis, apparently translated this into the reassuring statements that there was 'no danger' and that the scientific community assured him it was a 'favourable situation'. Go home, have a glass of wine, he reportedly said.

The succeeding events have the air of a scripted tragedy. At 11 p.m. on 5 April 2009 there was a strong shock, and families had to decide whether to stay indoors or spend the night out in the town squares – the traditional response to tremors. Families who heeded the apparently 'scientific' reassurances remained indoors, and 309 people were subsequently killed in their beds when the devastating earthquake struck at 3.30 the next morning, flattening many modern blocks of flats.

Six top Italian scientists and De Bernardinis were accused of manslaughter and stood trial in Italy. The scientists were not accused of failing to predict the earthquake, contrary to the impression given in some news reports in the UK, as it is acknowledged that this is currently impossible. The trial instead focused on what was communicated to the public. Did they appear, or did someone appear on their behalf, to predict that there would *not* be an earthquake, despite their own knowledge of the uncertainties? If so, they had not read the manual.

A key question is what scientists and officials think that people want to hear. One common refrain – often from scientists – is that people crave certainty and that that is unreasonable. The prosecution's claim in L'Aquila seems to have been, on the contrary, that the one thing they were sure they didn't want was to have *un*certainty suppressed. The prosecution has been caricatured in some quarters as typical of a country that tortured Galileo, typical of a public demand for fortune-telling from necessarily uncertain scientists. The real issue is arguably almost the reverse.

One witness, Guido Fioravanti, described how he had called his mother at about 11 o'clock on the night of the earthquake, after the first tremor.

'I remember the fear in her voice,' he said. 'On other occasions they would have fled but that night, with my father, they repeated to themselves what the risk commission had said. And they stayed.' His father was killed in the earthquake.

Another witness said: '[the messages from the commission meeting] may have in some way deprived us of the fear of earthquakes. The science, on this occasion, was dramatically superficial, and it betrayed the culture of prudence and good sense that our parents taught us on the basis of experience and of the wisdom of the previous generations.' Otherwise, he said, they'd have slept outside.* As it was, they stayed in.

All the accused were convicted by a lower court of multiple manslaughter and sentenced to terms of six years in prison but are expected, at the time of writing, to appeal. One of them pointed out that he had previously identified L'Aquila as the biggest earthquake risk in Italy. The quality of the buildings also had a lot to do with the death toll.

Who else might be culpable for issuing reassuring statements? The British TV weather forecaster Michael Fish jovially discounted the possibility of a hurricane in October 1987, and the subsequent storm killed 18 people. In 1990 the UK agriculture minister John Gummer was similarly reassuring about the safety of British beef when he force-fed his four-year-old daughter Cordelia a beefburger in front of the cameras. Over 100 people have subsequently died of variant CJD in the UK.

Of course, there's a difficult balance to be struck between reassurance and precaution. For every unheeded warning about a sub-prime crisis or cod depletion, there's an exaggerated claim about the potential dangers of saccharin or the Millennium Bug.

So although the Italian prosecution seems harsh, it's a salutary warning that people's emotions and intelligence should be treated with respect. People need full information and guidance for action, rather than just reassurance, and their concerns must be taken seriously.

---

* For a nuanced account of what went wrong in L'Aquila and what the scientists were actually accused of, see the article in *Nature* by Stephen Hall[10] and the below-the-line discussion that follows.

11

GIVING BIRTH

'UHH!' SHE SAID.
        Norm shuffled his notes ...
'Uhhhhhhghnnng!!'
... bit his lip.
'When will this *thing* be out?!'
'Erm, hold on,' he said, shuffling even faster. 'Right, yes, got it,' he
said. "Among 2,242 women with spontaneous onset of labour, the
median duration of labour for those delivered vaginally was 8.25 hours
in para 0, er ... 5.5 hours in para 1 and 4.75 hours in para 2+ mothers."
You're a para 0.'*

He smiled.
'Oh God!'
Norm had spent days, days compiling data.
'Another one ...'
He had footnotes on all his sources.
'Oh Goooodddddd!'
Everything written down alphabetically for ease of reference.
'Phu, phu, phu,' she breathed.
He had been able to tell her at critical moments how likely she was
to die.

---

* Meaning she's not had children before.

She grabbed his sleeve.

And had relative risk ratios *at his fingertips* for various anaesthetics.

'Tell me one more ... and I'll ... nggghhaaa ...'

Ah, threat of violence ... under 'V'. Here it was. Page 12. With the risk factor assigned to the psychological profile he'd done on her: 'Personal injury – low. Violent verbals – high (possibly of personal nature referencing anatomy)', he had written.

'You ... dick! You stupid, stupid f... nnnnnnggggghhh.'

Excellent, he thought, spot on with that one. But then he dropped his notes. This, of course, was about the time she moo-ed.

'Mmmmmnnnnnooooooooo!'*

Just as he was on the floor, scraping together the pages.

'Animal noise, animal ...,' he mumbled, but the pages were all over the place. Although now he thought about it he didn't recall anything about ...

'Nearly there,' said the midwife. 'You're doing brilliant ...'

Well, thought Norm, depends what metric she uses. Performance, P... P.... He should have used staples. Time-wise, his wife was about average, if truth be told. He was about to ask from under the instrument trolley what the midwife meant when a note caught his eye on a meta-analysis of randomised controlled trials of vitamin K injections for new-borns, he heard a gurgle, and the midwife tapped him on the back.

'Norm... Norm... It's a boy.'

And there he was.

---

* Accounts of childbirth in literature seem to have been mostly sanitised affairs until the 20th century, and mostly from a male point of view, like these from Kelvin and poor, befuddled Norm. Queen Victoria was unusually and famously forthright: 'I think much more of our being like a cow or a dog at such moments; when our poor nature becomes so very animal and unecstatic' (quoted by Helen Rappaport[1]). But at some point women writers made the subject their own and the writing became more explicit: 'the pain was no longer defined and separate from her but total, grasping, heating, bursting the whole of her, head, chest, wrought and pounded belly, so that animal sounds broke from it, grunts, incoherent grinding clamour, panting sighs' (A. S. Byatt, *Still Life*[2]). Or see Sylvia Plath, 'Metaphors' (1959): 'I'm a means, a stage, a cow in calf.'

'A beautiful baby boy,' she said.

And Mrs N with a huge sappy smile, the baby in her arms, the midwife grinning like a slice of melon.

But now, 'beautiful' ... A tomato thrown at a wall, more like. Always tricky these aesthetic metrics.

> Email from Prudence to Norm.
> **Attachments:** *Health and Safety Executive: New and Expectant Mothers at Work: A Guide for Employers.*
> Norm. Fantastic news. A precious time. Do be vigilant. Wise words in the attached HSE guide for new mums. See especially ...
> – lifting/carrying of heavy loads,
> – standing or sitting for long lengths of time,
> – exposure to infectious diseases,
> – work-related stress,
> – workstations and posture,
> Love to all. Take care. Px.

'The thing with the missus's caeso,' said Kelvin in the pub later, 'was they couldn't put it all back in, did I tell you? No, not the baby, you berk, the other stuff, hours, at it for hours they were, 'cause the caeso's the safe one innit, high-tech beep beep beep like bollocks it is, anyway whaddya mean, I said to him in the blue romper suit, you're not meant to have bits left over, how do you have bits left over it's not an Ikea cupboard is it, know what I mean? I mean for starters there's more room afterwards, *comprendez vouz*? How can it *not* fit? Bloody Tesco bag of emptied capacity still no room and still fat and wobblin' after how does that add up on volume, eh? Bloody Tardis backwards. But two hours, mate, bodge job I reckon. 'nother cigar? Surprised she didn't leak Guinness, not even as if yer even surplus caeso bits are *useful* or anything not like yer Ikea thingies them ... what do you call them ... thingies you can't put caeso bits in a jar in the garage, well you could, but... now, hey, that is what they do, innit, hospitals, cut you to bits and put half of them in a jar, specimens, pickled, come in for an op and half gets Branstoned, that's the scam, oh-dear-where's-that-go-can't-seem-to-ah-well-never-mind-Sir/

Madam-just-pops-them-in-a-jam-jar-shall-we?' Loses weight mind, good on a diet. More Scratchings? Which bits anyhow, biologolically speaking, you considered that? Curled up well-packed insides I suppose in their defence but wouldn't fit, nah! You believe that? Bodge job, I reckon, what are the chances Norm 'nother pint? Go on you Zulu warrior skull it!'

MOTHERHOOD IS the most natural thing in the world. Is it also one of the most dangerous? In 2010 around 287,000 women worldwide died giving birth, 1 in every 480, equal to 2,100 MicroMorts.[3] That's similar to the average dose of acute fatal risk incurred by a British citizen about every six years, condensed, usually, into a matter of hours, as Norm says.

Even so, how useful are those mortality figures? How useful is Norm in a delivery room? That is, how often is the decision to have children – for those who have a choice – a calculation of risk, even though it might be the riskiest thing a woman does, especially in developing countries?

Danger that makes little difference to our decisions is a good example of probability's real-life limitations. If you think Norm sounds absurd or irrelevant here, it is for a reason. If emotion or compulsion seems more important than data as Norm scrapes around under the instrument trolley, remember that about risk in general.

But only up to a point. For the numbers do tell extraordinary stories. They are also the focus of immense effort and concern. In some countries – Chad, Somalia – the risk is five times higher than the global average. Here the maternal death rate in childbirth is a frightening 1 per cent. This risk – 1 in 100 births, or 10,000 MicroMorts – might be considered the natural rate, in that it probably applied in all its brutality for millennia. It is one of the highest risks in this book, although it is eclipsed by some of the unnatural, historical risks somehow contrived by medical authorities. The *natural* rate of death in childbirth is by no means the worst.

For in the developed world, although maternal mortality has improved dramatically, the improvement has been slow, erratic and marked by terrible suffering. As any number of church memorials testifies, even among the upper classes childbirth was dangerous. About 150 years ago, 1 in 200 women still died this way in the UK, many from infection or 'puerperal fever'.

In charitable laying-in institutions the risk was higher still, worse even than the Stone-Age rate of 1 in 100. Giving birth in this period was safer at home, a fact recognised in 1841. Midwives were rightly feared, but were not nearly so deadly as doctors in some charitable hospitals. In Queen Charlotte's, London, 'which possesses a great reputation as a school of obstetric practice', an extraordinary 4 in every 100 mothers died. Giving birth in Queen Charlotte's was a risk equivalent to 40,000 MicroMorts, four times as bad as the natural rate of maternal mortality and far more dangerous even than a full year on active service in Afghanistan (around 5,000 MMs in 2011).[4] A hospital could be more treacherous for women giving birth than a modern war zone. And the enemy? Doctors and hospitals killed most of these mothers, not labour itself. If we compare their mortality rate to the natural rate, it implies that three out of four maternal deaths at these institutions were directly caused by medical professionals more deadly than the Taliban.

In Vienna in 1848, a Hungarian doctor, Ignaz Semmelweis, was an exception. He compared two clinics, one for training medical students and one run by midwives. The medical students had two or three times the midwives' rate of maternal mortality, averaging about 10 per cent, ten times the Stone-Age rate.

In one month, December 1842, Semmelweis reported 75 deaths from 239 births, for a truly incredible rate of almost one in three mothers dying during or after childbirth. The hospital was a slaughterhouse.

Eventually, he found out that the students – and their professors – went from handling corpses in autopsies to examining women in labour without thinking to wash their hands. He concluded that they were carrying 'cadaverous particles' and that it would be safer to give birth in the streets. Some women preferred to do just that if they went into labour on the day of a student clinic. Semmelweis instituted hand-washing with chlorine. The death rate dropped in one month from 18 per cent to 2 per cent.

But his genius proved unhealthy. He was dismissed, moved to Pest, ever more disturbed by the general dismissal of his opinions, and wrote

Figure 15: **Monthly mortality rates at Vienna General Hospital maternity clinic, 1841–9**[5]

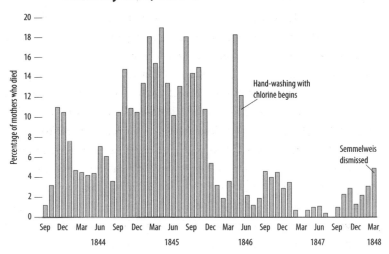

offensive letters to major obstetricians throughout Europe calling them murderers, with some justification. His behaviour became more erratic and embarrassing and in 1865 he was lured to an asylum, where he died two weeks later – by cruel coincidence from an infection after a beating from his guards. He was 47. It took another 30 years and thousands of unnecessary deaths for the germ theory to become established and Semmelweis to be vindicated.

And in the developed world today? According to the Office of National Statistics, each year in England and Wales around 50 women are registered as having died due to giving birth – around one a week.[6] The Confidential Enquiry into Maternal Deaths, published every three years since 1952 and now broadened into the Confidential Enquiry into Maternal and Child Health,[7] says that the true figure is higher – around double if we include deaths both directly due to the birth, such as from blood loss, and also indirect deaths, in which the birth made an existing problem worse.*

The Confidential Enquiry also found that around 150 children a year

---

* A helpful guide to terminology and statistics is given by patient.co.uk.[8]

in the UK are left motherless as a result of death from childbirth. The most common direct cause is still infection, as in Victorian times, but this now causes only 5 to 10 deaths a year in more than 700,000 maternities. The most common indirect cause is heart disease, with around 15–20 deaths a year: the stresses of pregnancy on a woman who had heart disease as a child has even featured as a storyline in the BBC radio series *The Archers*.

The UN reports 92 maternal deaths a year in the UK, to give an overall maternal mortality rate of around 1 in 9,000. By historical standards that is extraordinarily good, but still equal to about 120 MicroMorts – roughly the same as a motor-bike ride from London to Edinburgh and back.

Unlike in most other countries, in the UK the risk of childbirth has not fallen in the last 20 years. There is also, still, a strong social class gradient, with five times the risk in lower compared with professional classes. Older mothers are also at higher risk.

Taking the official UN data used for international comparisons, in Sweden the risk is only 40 MicroMorts, a third of that in the UK. As Semmelweis showed, cleanliness matters, and if you accompanied your wife as she gave birth in the US you would have to wear a gown, whereas in the UK you can wander in from the garden. Even so, the official maternal mortality rate in the US of 210 MicroMorts – more than five times that of Sweden – puts it level with Iran in the international league table.[9] Although, inevitably, these figures are disputed.

The American comedian Joan Rivers said that for her the ideal birth would be if they 'knock me out with the first pain, and wake me up when the hairdresser arrives', which is essentially what happened in Germany at the start of the 20th century with a form of anaesthesia known as 'twilight sleep'. Women could not even remember the birth. Anaesthesia had been popularised by Queen Victoria in the 1850s – before which time women had no choice but to suffer Eve's punishment, as described in Genesis: 'In sorrow thou shalt bring forth children.'

The fear of giving birth is called tokophobia. In the UK there is an organisation called the Birth Trauma Association, which says that anxiety is widespread and phobia not unusual, although in the extreme case

Figure 16: **Average MicroMorts for women giving birth**

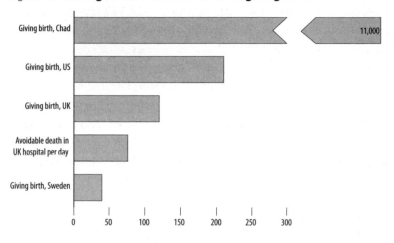

of phobia this seems to be expressed in terms of disgust as well as concern about safety.

That they think it worth trying to overcome even this degree of fear shows how childbirth, in particular, presents a difficulty for any simple calculation of harm, namely: what's it worth? Because there is also, obviously, a benefit, a baby. The fact that millions of women and men accept the risks even if they have access to contraception and no qualms about using it (although see Chapter 8, on sex) shows that even severe risks of harm might be worth it.

In fact, there's evidence that people lower their estimation of the risk the more benefit they expect. This means not simply that for some people the benefit outweighs the risk and for others it doesn't, but that those who think the benefits are large also tend to think that the risk is *objectively* lower for themselves and everyone else. An imperfect rule of thumb might be: the more you expect, the less you worry. But why should the good reduce the bad? Compensate for, yes, but reduce? The psychologist of risk Paul Slovic calls this the 'affect heuristic'. If you like an idea, you find it harder to see how it might hurt you.

Would it be safer to press for a caesarean section, as Kelvin's wife did? It is almost certainly a myth that Julius Caesar was born by caesarean, since it was then only used when the mother was dead or dying, and

Caesar's mother seems to have been still alive when he was grown up. The first woman to survive a caesarean was said to be a 16th-century wife of a man skilled in neutering sows, which supposedly gave him intimate anatomical knowledge. Regardless of the myth, Caesar's descendants are enthusiastic about the practice, since now nearly half of births in Rome are by caesarean, rising to 80 per cent in private clinics. At 170 Micro-Morts[10] the official record of deaths during caesarean section appears to suggest that a caesarean roughly increases the risk by a half. But the risks are hotly contested, particularly as it is difficult to separate the risks of the operation itself from the risks associated with the reasons for having a caesarean in the first place.

Of course, all sorts of things can go wrong that don't lead to the death of the mother but can nonetheless have a deep impact. The most common is post-natal depression, which affects around 10 to 15 per cent of new mothers. It's another risk that runs back into the problem of how hard it can be to see past the benefits to the risks, especially when the benefits are large, or to take seriously the chance that our own feelings are a poor guide to how we will feel in future. 'I know that I want a baby, that I want it very much, that I feel happy at the thought of it, that the thought of not having a baby makes me unhappy and that therefore having a baby will make me happy.' Who is to tell you that you are wrong? But the risk of depression is not small. We can try to persuade people to take this risk more seriously, and perhaps we should, not least by offering proper help after the birth if problems arise. But we should not expect it to change people's hopes and behaviour.

12

GAMBLING

NORM'S GAMBLING PHASE began after an unprecedented streak of rollovers left the National Lottery Jackpot on £14 million – and he calculated that £14 million would buy every possible lotto combination. So if he shelled out £14 million in advance on tickets, winning would be a cert.

Plus that all the lower prizes he'd also clean up would yield a hefty profit. At which his eyes had an unusual – for Norm – Rambo look.

'Gosh,' he said.

Though when he asked at breakfast how to raise £14 million, his wife didn't seem to hear.

'I said ...'

'I know ...'

'But I've done the calculations,' said Norm, 'it's a rational strategy.'

'I think you may have missed the point of gambling,' she said.

'It's a sure thing.'

'As I said, I think you ...'

'Winning?'

Of course, there was the small practical problem of how to buy 14 million tickets. At about 30 seconds per, that was still nearly 7,000 days at the machine even full-time in Mr Singh's. But what finally put a spanner in the plan was when he realised he had overlooked something.*

---

* Guess the flaw in Norm's plan. Then see the discussion in the second part of this chapter.

Which is when his interest ended. No rational point, given the odds.

HOW IT ALL BEGAN, years previously, at a slot machine at a family wedding reception for the lad who was now trying to sell his corn flakes to Kelvin outside Ladbrokes:

'Oh, my lad! What a lad!'

'Play one for me will ya, sunshine?'

'He has the luck!'

'He's the knack!'

'Blond hair and blue eyes, it's angel luck!'

'Here, here's another, little fella, go on, son, go on.'

And the gorgeous lights flashed, and the coins pumped and cheered.

How it ended:

'For 50p then? Cost a pound, mate, come on. More than that, look: £1.23, it's on the box, come on, 50p!'

'I don't like corn flakes.'

'What then? The milk? The lot – a fiver the lot? Come on. Please. A fiver. Lend us. I'll win it back. I'm due. Just a fiver, mate!'

And then, because the fare he'd kept safe in his back pocket hadn't been safe at all and eventually went the same way as the shopping, and there was the long walk home to find the note saying that she'd left after too many lies, and because it was a grand, more, blown in two days, a grand in her name they didn't have; and because he'd tried, God he'd tried, he'd even bought half the weekly shop first to be safe and then sold it for F-all because the perfect moment to stop never came, there never was a win big enough; and because now his plastic is no good and he's crying again, and he sleeps in his clothes; and because his dad has already bailed him out but he blew the bail-out and just the thought of that carves another hole in his soul, so back comes the sickness and self-loathing as he curses at the loser in the mirror thinking everything he has is nothing and he is 24; because he is, in the end, lost; because of all this, but still somehow holding on to the wreckage of a daydream, he calls the Gamblers Anonymous free helpline sobbing in a piss-stinking phone box. Which was sad, because this was what he lived for.

But Kelvin didn't want the guy's corn flakes. He wanted Cheese Rind at 4–1 to get its snout in front on the big screen in the 7.30 Platinum Stakes at Reading, and as the money rode the track and the dogs strained, he was beside himself with exhilaration. Norm said he was a fool, given the probabilities. Only the next day, after Snow Queen topped the evening at an unlikely 7–1 over the 660-metre hurdles with fine late speed on the stretch, he won £81,281.52 on an accumulator. It was getting on for enough to buy a Maserati – if only he hadn't broken all his rules come the weekend and poured 40 grand down a women's tennis match on a Russian he'd never heard of because someone said Russians were good but this was a donkey in a tutu. So? He'd won it first, which meant he lost nothing, and decided to go instead for either a Porsche or BMW 335i convertible. Naturally, he flipped for it, bought the Porsche, put it in a canal, pissed that night having gambled on third-party insurance.

> 'Then said Jesus, Father, forgive them; for they know not what
> they do. And they parted his clothing, and cast lots.' Luke 23:34[1]

We can't know when people first started harnessing unpredictability as a fair means of making tricky decisions. But what went for Christ's clothing continues with some school admissions, selection of juries or deciding which survivor to eat in the lifeboat,* or sometimes buying a car.

Young men sent to Vietnam were chosen by drawing from a box of capsules containing birthdays. Unfortunately in 1969 they put the capsules in the box in order of month but didn't mix them well, with the result that it was bad luck to be born later in the year: 26 out of 31 birthdays in December ended up drafted.[3]

The idea of a lottery should be to give everyone an equal chance. This randomness allows the intervention of whatever belief one has in God, fate, luck or fortune, a process still seen in the bizarre rituals and mascots used in casinos and competitions. Norm, by trying to beat the lottery, is

---

* See, for example, *The Luck of the Draw: The Role of Lotteries in Decision Making*, by Peter Stone, which argues – with others – that the perceived problem with using lotteries for social decision-making is that they actively prevent reasons from playing a part in the decision, but that this very quality has a 'sanitizing effect'.[2]

up to his old tricks again of trying not just to tame chance but to make it irrelevant. Now that would be power. But would it be fun? Would it even be interesting? Maybe that's what Mrs N is getting at. If you deny life's lottery, do you stifle life?

Devices that embody randomness have been used for leisure for at least 5,000 years. The ancient Egyptians, of a long winter's evening presumably, sat around and played board games in which moves were decided by throwing an 'astragalus', a bone in the heel that can fall on one of four sides. If you buy a leg of lamb you can, with a little effort and mess, extract your own astragalus and use it for games or divination, just as the Greeks and Romans also did. Then people started putting money on the outcomes of the throws, and the move from game-playing to gaming began,[4] using either astragali or dice: indeed modern versions of the Bible now assert that 'the soldiers gambled for his clothes by throwing dice'.[5]

Gambling became so popular that the Romans tried to restrict it to Saturdays, but even the Emperor Claudius played obsessively and wrote a book: *How to Win at Dice*. People continued to make bets and quote odds: in Paris in 1588 you could get 5 to 1 against the Spanish Armada sailing to invade England, although this was probably a ruse by the Spanish.[6] But what is remarkable is that in all this time, right until the Renaissance in the 1500s, nobody analysed gambles mathematically. Maybe they thought the outcomes were decided by some external force of fate, but it's now thought that the gap between theory and practice was too great, and the idea of putting a number on chance – the probability – just did not occur.[7]

The first book that started to work out odds was written in 1525 by an Italian, Girolamo Cardano, another obsessive gambler with no time for superstition.[8] He came up with the idea of counting the favourable outcomes – say the six ways of throwing 7 with two dice – and dividing it by the total of 36 possible outcomes, to find a chance of one in six of throwing a 7 with two dice. This seems obvious now, but it was a remarkable achievement. Although he got a lot wrong too. He seemed to think that, just because an astragalus can land on one of four sides, each side is equally likely; but a brief experiment shows this is not the case (try it – astragali are uneven). He also thought that three throws of a die would be sufficient to give a 50:50 chance of a particular face, say a 6, a mistake that could have lost him a lot of money. You can try this as well. (The odds would be 50:50 only if there were no repeats.)

The Chevalier de Méré, a more perceptive gambler, in Paris in the 1650s, reckoned from his gaming that if he bet that he could throw a 6 in four throws, then the odds were slightly in his favour. Whereas if he threw two dice and bet he could land a double 6 in 24 throws, then the odds were slightly against him. By chance (or fate) he brought the problem to the attention of two of the smartest mathematicians at that or any other time, Blaise Pascal and Pierre de Fermat (he of the Last Theorem), who confirmed that he had a 52 per cent chance of winning the first bet, and a 49 forty-nine chance of winning the second,[9] so the Chevalier had actually got it right, from what must have been extensive and expensive experience. The Chevalier also presented the mathematicians with the 'problem of points' – if a game has to stop before the end, in what proportions should the stake be divided up? A modern counterpart is the Duckworth-Lewis method for allocating runs when a cricket match is stopped, this time designed by statisticians – and so suitably incomprehensible.

The scientific assessment of odds then disappears for a few centuries, as the 1700s became a golden age of betting on the basis of gut feeling rather than calculation. It was the time of 'eccentric wagers', such as a large sum won by the Count de Buckeburg in 1735 for riding a horse from London to York seated backwards.[10] There were also huge bets on cricket matches, and a predictable consequence was the match-fixing scandals that erupted in the early 1800s, with spectators at Lord's bewildered at the sight of two nobbled teams both desperately trying to lose.[11]

As a result, bookmakers were banned from Lord's in 1817, lotteries banned in 1826, and the Gaming Act of 1845 made gambling debts unenforceable by law. Cricket was turned into the archetypal gentlemen's game.

That is, until recently, when heavy gambling led to match-fixing and, owing to the ability to bet on every detail of the match, players were bribed to mess up specific balls – an example of spot-fixing.[12]

Victorian morality abhorred gambling, and only with the liberalisation of the 1960s did it begin to become a respectable pastime. Now, according to Gamble Aware,[13] 73 per cent of adults in Britain gamble each year. Even excluding the lottery, the figure is still almost half the adult population. The Gambling Commission estimates that they lost around £6 billion in 2009/10 – not counting the lottery – that's an average of £100 per man, woman and child, or around £200 per gambler. And that's the average. Most people don't lose anything like that, so some must be pouring it away.

Two statisticians once wrote a dense textbook on probability theory that they called *How to Gamble if You Must*, which led to many disappointed purchasers and a subsequent change in title to the more accurate *Inequalities for Stochastic Processes*.[14] But what would be a good gamble now, if you must? We are not talking about corporate gambling on the financial markets, or games of skill such as poker, or the sports-betting organisations employing PhD mathematicians and using sophisticated statistical models. We mean gambles for the fairly naive punter.

Lotteries prefer not to think of themselves as gambling, and the UK National Lottery does not even come under the Gambling Commission, but nevertheless takes in nearly £600 million each year. The jackpot in the main draw is won by matching 6 numbers drawn from 49, and as there are around 14 million possible combinations that's a 1-in-14-million chance of winning from a single ticket. That's also about the chance of a 50-year-old woman dying from any cause in 15 minutes, or about the chance of being knocked off your bicycle and killed during a 1-mile journey, or DS's chance of having a heart attack or stroke in the next 7 minutes. – a bit more than the time it takes to buy a ticket. But with around 30 million ticket sales for each Saturday draw, on average 2 people should win each time. And the same number will die within a few minutes of buying their ticket. But nobody blames the lottery – though they would if it had been a vaccine.

If nobody wins, the big prize is rolled over. If this carries on, the jackpot could grow so big as to be more than the cost of buying every combination possible, say if the UK jackpot ever reached more than £14 million. It would then become an interesting business proposition – as

Norm realises – to buy all possible tickets, and although this strategy would be difficult to organise, it would win all the subsidiary prizes as well.

This has happened. In 1992 the jackpot in the Irish lottery stood at £1.7 million in spite of it costing only £973,896 to buy all the tickets and so be sure of winning. A Dublin syndicate managed to buy 80 per cent of the tickets despite the efforts of the Lottery organisers to stop them. The flaw in Norm's plan is that, although they won the jackpot, they had to share it with two other punters, and so would have made a loss, but with all the lesser prizes managed to make a small profit. Norm knows that winning is a cert – but not who else might win too. That's still a lottery.

That's the problem with this idea: massive payouts attract vast ticket sales, increasing the chance of having to share the big money. Take the US Mega-millions $1 lottery: in March 2012 the pre-tax jackpot stood at $656 million, although there were only 176 million possible tickets. In the end there were three jackpot winners – it would have been a high-risk gamble to try and buy all the tickets.

But even if a loss is made, lotteries present an obvious opportunity for money-laundering, and the World Lottery Association even produces a guide on how operators can guard against this.[15]

If you want to have the best betting odds, then visit one of the 150 or so casinos in the UK. You will still probably lose, but a European roulette table has only one zero, when the bank takes all the bets (an American table has two zeros), giving the house a small edge of 2.7 per cent, meaning the casino returns on average 97.3 per cent of money staked on roulette. This compares to the 45 per cent return of the National Lottery, which doesn't sound great but is better than other lotteries.

Betting on horse-racing and other sports is still a mainstay of the 8,000 or so betting shops in the UK, which operate at around a 88 per cent payout, although Fixed Odds Betting Terminals (FOBTs) playing roulette at a 97.3 per cent payout are now more popular than the racing and are increasingly accused of encouraging problem gambling. FOBTs have a high payout, but the playing is so rapid and compulsive (DS can vouch for this) that they are staggeringly profitable – UK betting shops are only allowed four per shop, so outlets are proliferating in High Streets just to house the FOBTs. But these don't have the drama of an accumulator bet, in which a small initial stake accumulates over a whole series of connected bets provided they are all successful, such as picking 19 correct football results in a row and winning £585,000 for an initial 86p stake.[16]

Gambling is increasingly becoming a private activity conducted at home on the Internet – in 2008 5.6 per cent (1 in 18) of the adult population played online (non-lottery) games. These websites may provide a link, hidden at the bottom of the page, to Gamble Aware and Gamblers Anonymous, but problem gambling is estimated to afflict around 1 or 3 per cent of adults, depending on who you listen to. 'Pathological gambling' has been considered a psychiatric disorder for many years, with an official diagnosis if you are someone who ticks at least 5 of the following boxes:[17]

- is preoccupied with gambling (e.g., preoccupied with reliving past gambling experiences, handicapping or planning the next venture, or thinking of ways to get money with which to gamble)
- needs to gamble with increasing amounts of money in order to achieve the desired excitement
- has repeated unsuccessful efforts to control, cut back, or stop gambling
- is restless or irritable when attempting to cut down or stop gambling
- gambles as a way of escaping from problems or of relieving a dysphoric mood (e.g., feelings of helplessness, guilt, anxiety, depression)
- after losing money gambling, often returns another day to get even ('chasing' one's losses)
- lies to family members, therapist, or others to conceal the extent of involvement with gambling
- has committed illegal acts such as forgery, fraud, theft or embezzlement to finance gambling
- has jeopardised or lost a significant relationship, job, or educational or career opportunity because of gambling

- relies on others to provide money to relieve a desperate financial situation caused by gambling

These terse statements cover a pile of misery. So why do people continue to gamble in games of pure chance, when they know the house wins on average? The problem is that, although we may rationally know it's all just chance, and nothing we can know or do can affect the odds, we still tend to think that a good result is 'due', or that really we have some control over what happens, and that 'near-misses' are as important as they (truly) are in games of skill such as football.

There is only one NHS clinic for problem gambling, appropriately situated in Soho.[18] Soon problem gambling will be officially considered an addiction, similar to drink or drugs. And if such behaviour is to be medically labelled as an addiction, what next? Obsessive shopping?

But do people really gamble as a rational substitution for buying a pension? If they do, they are truly irrational. But if we are not doing it as an investment strategy, and we don't wreck our health and family, perhaps it's not such a bad thing. In that case, gambling can be, well, fun.

So let's say you wanted to have some, and fancied a £100,000 Maserati, like Kelvin, but sadly you had only a pound. And let's assume you are a cool, rational customer who wants the best odds (admittedly this is an implausible combination of characteristics, which is why Norm might be missing the point of gambling). If you buy a single lottery ticket, and if your choice of 6 numbers matches 5 winning balls plus the bonus number (a 7th ball drawn) then this generally wins around £100,000 and has a probability 1 in 2,330,636.

Or you could go for an accumulator on the horses or dogs like Kelvin: pick a meeting with 6 races, and in each race choose a horse at medium odds of around 6–1. An accumulator, in which the winnings of each race are passed to the next horse, will give you $7 \times 7 \times 7 \times 7 \times 7 \times 7$ = £117,000 if they all win. Given a bookmaker's margin of, say, 12 per cent each bet, the true odds may be around 1 in 230,000, ten times as good as the lottery.

If you can find a casino that will let you bet just £1, place it on your lucky number between 1 and 36. When it wins, either leave the £36 there or move it to another number. When that comes up too, move the £1,296 you now have to another number, or leave it where it is – it doesn't make any difference to the odds, but somehow it seems that the chance increases when the money is moved. When that comes up, you will have £46,656, so move it all to Red, and when that comes up you will have £93,312, almost enough for your Maserati. The chances of this happening, on a European roulette wheel with one zero, are $1/37 \times 1/37 \times 1/37 \times 18/37 = 1$ in 104,120, twice as good as the horses.

So roulette easily gives the best chance of that shiny Maserati for a quid. Perhaps it's better to start saving.

13
AVERAGE RISKS

Norm's life lacked ... what? He was 38 years old and didn't know. But he knew missing when he saw it. This depressed him, although in truth not much. He was more moderately pissed off than depressed proper. That almost depressed him too, in a middling kind of way. Where was life's, erm ... you know? He ripped back on his chair and considered the curtains.

Like any average kind of guy, Norm knew in his heart that he was better than most. Proving it, that was the problem. In an effort at least to stand out he had taken up groovy habits like wearing striped socks or, for edge – you need edge – swearing a bit. Taking pleasure where he liked, whatever anyone said, he slipped into his shopping basket – alongside a two-pint carton of semi-skimmed milk, pre-packed sliced ham, breakfast cereal and a chicken korma (mild) ready-meal – a bar of milk chocolate.[1] But that's how Norm was these days: more of his own man, picking out his own personal style in the Boden catalogue. But still he lacked ... you know ...

He turned over an old envelope and wrote 'NORM' at the top, underlined, twice. On the left, he wrote 'Earnings', drew a line down the middle and scribbled on the right: '£28,270'. He stared at the number.

It was weirdly familiar. He looked it up. He was right. It was the UK average.[2] How funny, to earn exactly the average for a full-time male employee.

'Height', he wrote, and then '5'9"'. He looked that up too, on the Office for National Statistics website. Also about average. Not too surprising, that one, he supposed.

Weight: Just over 13 stone. He stared at the number, too, thinking. Then likewise looked it up. Again, average.

Weekly working hours: 39. Which turned out with a little research ... he shifted in his chair and chewed his pencil.

Age at marriage ... number of cups of coffee drunk per day ... this was getting more than a little weird.

As he scribbled one stat after another on the envelope and then chewed the pencil as he typed a search into Google, they all gave the same, uncanny answer.

Age at which he had his first child ... He was afraid to find out. Commuting time, hours spent watching TV, shoe size, number of fillings? He hated to think, but felt the answer in his bones already. How could one man be able to tick so many boxes, all in the middle?

For years Norm had yearned for the big event, the moment that would mark him out as unique, the way people do. He might even have worked harder for it but for the effort. Instead, he raised himself in his estimation by fine observations of the inferiority of the world at large: TV presenters' grammar, other motorists' speed – damn eejits who drove faster, doddering gits who didn't. And all the while, here he was, bang in the middle. The implications were terrifying. He'd have to read the *Daily Mail*. He did read the *Daily Mail* (his wife's).

Could he fight it? He was tempted. Yes, to rebel, to throw off what seemed a fate of stifling, mediocre mediocrity, hyper-normality, and do something uniquely impulsive and, oh, anything really ... like, like ... getting drunk.

Norm tapped his pencil on his teeth and stared again at the curtains. Was he humiliated? He wasn't sure. It was a lot to take in, this strange new status as some sort of, erm ... He tapped his teeth some more. Then, abruptly, he stopped.

He sat back. He smiled. He put his hands behind his head. His smile grew, like that of a man who knew. He radiated satisfaction.

'Oh yes,' he said, not unlike John Major. 'Oh yes!'

THIS IS NORM'S APOTHEOSIS, the shining pinnacle of his narrative arc and moment of self-discovery – how bog-standard average he is. Quite simply, it makes him unique. That sounds absurd. How can anyone be uniquely average?

Norm can, because the average is not normally an individual quality, it describes everyone blended together. So it might apply to no one in particular, or at least to no individual, except, uniquely, to Norm.

He was always an average kind of guy – that much we knew – but we had no idea what a paragon of the ordinary he was. Nor did he. Plumb in the middle of the middle, the model of mediocrity, he probably drives a Ford Fiesta and holidays in Ibiza. Nothing wrong with that, of course, but is it a special cause for satisfaction? And yet Norm smiles.

Perhaps because being average is strangely useful if you worry about the future, especially if you look to numbers for help. For all risks on which we stick numbers and probabilities are, in fact, averages.

So when it is said that you have a 20 per cent higher chance of developing colorectal or pancreatic cancer if you eat an extra sausage every day, it does not mean that you personally face this increased risk; it means the average person does. So the average risk – or simply risk, for all risks are averaged over some group or other – describes Norm's future more reliably than it describes anyone else's, all those others – like you – who in some particular way differ from the average. Are they talking about me? On the whole, no. In Norm's case, yes. It's a bit like the child's fantasy that the world is designed around me. For Norm, it's sort of true. What an ego trip for a man who can't even make up his mind about Marmite.

Norm – in a neat paradox for a man so unexceptional – is the archetype for man. He is no one and everyman. He is in no way outstanding and, precisely because of that, outstanding. He is both prototypical and a one-off (add more oxymorons at will ...). They should put up his statue – outside Tesco.

Not that this idea is without problems. We'll come to those in a moment. Meanwhile, allow Norm his strange glory.

The idea of the average person has itself become ordinary, but it is a statistical invention only about 150 years old. It originated with a

19th-century Belgian statistician, Adolphe Quetelet, who believed that the essential characteristics of the average man, '*l'homme moyen*', could be discovered by gathering data about the whole population, putting it on a graph and looking for underlying patterns, peaks or regularities.

Quetelet wrote: 'If an individual at any given epoch of society possessed all the qualities of an average man, he would represent all that is great, good and beautiful.'*

Cue Norm. As Mr Average, he has more hope than most of being in tune with life's regularities. Though if he had entirely average characteristics, he would also have about one testicle and about one breast. That is what happens when you add together a whole population. You find that an equal share of all the parts does not necessarily add up to a coherent human being. But Quetelet – a brilliant statistician – was not, in principle, deterred. He seems to have believed that the average was more than an abstraction, and thought that many averages represented genuine physical or mental capacities awaiting discovery, including moral capacities.

But Norm's satisfaction is everyone else's problem. Ordinarily, as we say, no one is precisely average in the many respects that determine the risks they face. Perhaps they are a little heavier, a little richer, or poorer, slightly more tense, sleep a little worse, are taller, slower, more sedentary, more tempted by cake, picked up some odd genes from a deranged ancestor, and who knows what else that might make a difference to their future, might tip the balance for or against survival to make the odds better or worse than average.

And there are other problems with defining life's prospects by the average. Some averages are ridiculous. Probably no one has the average number of feet. In fact, not even Norm can be average in every respect without running into a few logical absurdities. For example, he cannot be the average age all his life.

He is also unlikely to be the average weight for a man and the average weight for a 31-year-old unless by luck these turn out to be the same.

---

* Quetelet derived *l'homme moyen* from the mean values of measured variables, which generally followed a normal distribution.

Different categories have different averages, and we all occupy many categories. Some averages are mutually incompatible. In other words, Norm cannot really be the average man, but can only ever be the average for some subset of men, sometimes a small subset, and by choosing one subset he ceases to be average for others. Very often the true average cannot exist as a man, or woman, at all. This points to a difficulty about all risk. It often describes the dangers that apply to someone who isn't there.

All this would be a terrible thing to say to Norm, so let's not break it to him. In any event, even if the average is theoretically imperfect and a bad fit for the complexity of life, it might work well enough to give Norm a practical steer as he decides what to do. We will see.

What indisputably does exist is variation around the average. Most people are not average, we all deviate from the norm/Norm, and these random deviations from Quetelet's *homme moyen* contain more of the real grit of life. They also change every individual's expectation of risk.

One of the best examples of bucking the kind of average that seems bound to determine life is the story of the American palaeontologist Stephen Jay Gould, diagnosed with abdominal mesothelioma at the height of a brilliant career with two young children, told that it was incurable and that the median* survival time after discovery was eight months. We might say that the risk – the average risk – of abdominal mesothelioma is death in eight months.

But this is not everyone's fate. It is the point by which half of all people who have it have died. And for those who haven't died by then, it is not the mid-point either. Gould lived for another 20 years and eventually died from an unrelated cancer. Averages of any kind, as Gould wrote in an essay describing this brush with mortality, are not immutable entities but an abstraction, and the true reality is 'in our actual world of shadings, variation and continua'.†

---

* Means and medians: for height, line up everyone from tallest to smallest and the median is the one in the middle; the mean is more like an equal share of all the height (i.e., what you get if you add up all the heights and divide the answer by the number of people).

† Gould tells his own story in a short, remarkable essay: 'The median is not the

There are shadings, variation and continua for all risks as for all averages. On average, men are taller than women. But then there is the Formula 1 boss Bernie Ecclestone, at about 5'3", and his former wife Slavica, at 6'2". There are, similarly, plenty of people out there who are Bernie Ecclestones of risk.

In fact, the average is sometimes misleading, not just about the odd individual but for the majority. About two-thirds of people in the UK have less than the average income (average in the sense of the mean). If we arranged the world's population in line, according to wealth, the average (mean) person would be about three-quarters of the way along the line.

Even MicroMorts and MicroLives cannot escape the problem of being averages. The average 10-MicroMort risk from sky-diving arises mainly from the deaths of obsessive sky-divers who do increasingly risky things. That is, the deaths are nearly all among experienced jumpers. The novice charity-supporting tandem parachuter may be taking just as much risk by simply getting drunk and walking home (but they probably wouldn't get the sponsorship).

Quetelet, no fool, knew all about this. He was every bit as sensitive to variations around the average. As he sat labouring over those reams of 19th-century data, the huge variation in human experience would have been obvious. The Body Mass Index, or BMI, by which we judge whether people are over or underweight, as well as normal, is also known as the Quetelet index.

So Quetelet stood between two ideas: one was the great scope of human variability; the other was the peculiar manner in which this variety seemed to contain an essence. As we will see in Chapter 14, on chance, the essence, the average, can be scarily predictable, but only at the right scale. This is the scale of whole populations, boiled down and their essence extracted. The problem is simply that this is not the scale on which individuals in all their variability live. Except for Norm. For the rest of us, when it comes to risk, none of us can attain the same state of Brahman-like self-knowledge as our hero.

---

message'.[3] For a longer, easy discussion of averages see *The Tiger that Isn't*, by Blastland and Dilnot,[4] or, in more detail, *The Flaw of Averages*, by Sam Savage.[5]

Norm is a quintessence, even if sometimes a logically absurd one. Not everyone can say that. In fact, no one can say that. But is this enough for even Norm to be able to navigate safely through a world of hazards, using average numbers? Or does it mean that risk is never really his risk, or anyone else's, and even he is a fool to think otherwise?

CHANCE

TALL, CHISELLED AND GORGEOUS under long, dark hair, Kelvin Kevlin's older brother Kevin shared the same love of chance. Except that for Kevin, Professor of Social Cognition at the Sorbonne and pop-up thinking-crumpet TV pundit, now visiting Oxford to deliver the celebrated Ronald McDonald public lectures, chance smiled on him.

He had a certain reputation for show-off controversy confirmed by his recent book *God/I*, and it was this that drew the crowds. At the first lecture a mathematician had been restrained from 'punching the F...reudian lights out' of a psychologist. At the second, a Nobel laureate theoretical physicist working on string theory spat from the gallery while someone tore the dust-cover from the upright piano and banged out some Wagner.

At the third and last, the crowd bustled. The talk, entitled: 'I Am Not a Piano Key', was billed as an assault on reason. Rumour had it the professor would urge supporters to smash their cars into the walls of Balliol College to prove they were alive, even if it meant death. Kevin, who spoke with fierce urgency and tended to jump straight into his lectures, pushed a strand of hair behind the ear, where a gold stud twinkled, and stepped up to the lectern.

'Reason is an excellent thing,' he said, scanning the faces, 'there's no disputing that. But reason is nothing but reason and satisfies only the rational side of human nature. The whole human life must include all the impulses, the will and the passions. It is not about extracting square roots.'

Was it madness, genius or fraud that danced in those eyes? He leaned on the lectern and glared at a front row of gigawatts of cerebral power.

'What does reason know? Reason only knows what it has learned – and some things it will never learn – while human nature acts as a whole, with everything that is in it, consciously or unconsciously. Even if it goes wrong ... it lives,' and he thumped his chest.

'Some of you look at me with pity; you think that an enlightened man cannot consciously desire anything bad for him. But *I* can. I might deliberately want what is bad for me, what is stupid, very stupid – simply in order to have the right to desire what is stupid and screw the obligation to desire only what is sensible. For this very stupid thing, this caprice of ours, preserves for us what is most precious and most important – this is our personality, our individuality.'

Some in the crowd cheered, or jeered, it was hard to tell. The Professor barely paused. Up in the gallery above the podium, the shouting was louder. A fight? Kevin thrust on.

'Give me every earthly blessing, a sea of happiness with nothing but bubbles of bliss on the surface, and even then out of sheer ingratitude, sheer spite, I say "risk it all!" simply to introduce a fatal fantastic element, simply in order to prove to ourselves that we are still people and not the keys of a piano, which the laws of physics threaten to control so completely that soon we will be able to desire nothing but by the calendar.'

By now, his hands were a conductor's, flicking, sweeping the air, his blue eyes enraged, his hair skipping, his voice rising.

'And that is not all: for not even the laws of physics can stop the play of chance that gives us limitless possibility and may even break the piano, that gives us freedom even to choose unreason. And even if we really were nothing but piano keys, even if this were proved by science and mathematics, even then we would not be reasonable but would do something stupid out of simple ingratitude, contrive destruction and chaos, only to make our point and convince ourselves that we are human and not a piano key!'

The gallery was heaving. A small scrum of young men moved towards the railing, clearing the crowd and pushing or pulling something heavy.

'If you say that all this, too, can be calculated and tabulated – chaos and darkness and curses – then people would go mad on purpose just to be rid of reason and make the point! For the whole of life is nothing but proving to ourselves every minute that we are human and not a piano key!'

In the gallery they bent down and lifted one end of the heavy thing. Its dark rectangular side appeared, poised, over the railing directly above the Professor's head. There were shouts, a panic in the front rows of the standing crowd, a scramble, as the upright piano jutted further over the railing, some awkwardness as it caught, was heaved again, tilted, and then, to gasps and screams, reached its tipping point.

The professor did not look up. He ignored the noise. He poured out his argument. Reason fought against reason, climbing over itself into a hymn of hysterical praise to human impulse as the slow turn of the piano hastened, its great weight released into a drop, and its dark mass fell.

For a moment after the crash there was silence, except for the hum of piano wire and a fluttering of the Professor's written notes in a haze of dust over a mound of splintered wood and the bones of the dark thing's metal frame. Then, with the diabolical aura of the all-time lucky bastard, Professor Kevin swept back his hair, placed a foot on the debris, leaned at the retreated, gawping audience – and screamed with laughter.

A day later he was accused by a group of Oxford academics of shameless plagiarism of Dostoevsky, an accusation he mocked on the grounds that if it was shameless it was hardly plagiarism, as plagiarism tries to hide its tracks whereas his was so blatant it could only be a tribute.* Also accused by the police of incitement to criminal behaviour, he argued on the basis of the definitive critical appraisal of Dostoevsky's *Notes from*

---

* The Underground Man, from whom Kevin nicks most of his speech, argues for the freedom to reject 2+2 =4. 'I admit that twice two makes four is an excellent thing, but if we are to give everything its due, twice two makes five is sometimes a very charming thing too.'[1] Other Dostoevsky characters rebel in similar ways, in order to affirm their individuality.

the Underground – a rant against rational egotism – that the argument therein was in fact parody and no incitement was intended or for that matter understood, since no one did in fact smash into the walls of Balliol. He was let off with a caution. He refused to assist the police inquiry into his attempted murder by a group of medical students in the Balliol First XV, an accusation later reduced to a charge of criminal damage, although the suspicion persisted that the whole incident had been contrived. A day later he resigned, without reason, he said.

NONE OF US KNOWS what will happen tomorrow, let alone years hence, and many don't want to. Like Professor Kevin. In a fine family tradition he prefers life instinctive and messy. Choices are not calculable anyway, so why try?

Others – like Prudence – want control over their future and take every care to get it, if they can.

Key to both Kevin's hopes and Pru's fears is chance. He loves its power to throw up the unexpected. She hates it for the same reason. Chance is the rogue that threatens either to wreck their best-laid plans or, as if by magic, pull a rabbit from the hat. (Norm, standing between them, thinks he can play the odds – to bet whether rabbit or wreckage is more likely.)

But what is chance? Philosophers have wrestled with the problem for centuries, from declaring chance all-powerful to wondering if it exists.

We're going to put the question in an unusual way with a dark edge of practicality, by asking: why does the piano miss Kevin? It wasn't meant to. He was all but a dead man. By what means did he survive?

Kevin himself answers by linking chance with free will as two expressions of the glorious muddle in life that makes us humans and not machines. He sets this against an idea of reason that he seems to equate with soulless determinism. For while strict necessity of cause and effect implies only one possible future, chance and free will in their different ways can turn away from this necessity.* In Kevin's eyes, then, the piano

---

* Though whether chance necessarily implies free will is a long argument. Some philosophers say that both chance and determinism are incompatible with free will, since they mean that life is either determined or accidental and neither

misses him because all of life, material and human, must have a 'fantasti-cal element' that breaks the chain of strict causality. This is the way he thinks life ought to be – and the way it ought to be lived – even if he betrays a doubt or two about whether that's how it truly is.

The dictionary is not so dramatic, defining chance mainly as a prosaic possibility: what chance of rain? In everyday use we often go further and give chance an improbable twist, saying that to 'take your chances', for example, is to be something like a gambler, a risk-taker who knows the odds are against success, but what the heck? If you wanted a story about chance, someone might tell an improbable tale about an unintended event on which the whole plot hinged, a strange coincidence or turn-up, the more fantastic the better.

And if chance brings calamity, when star-crossed Shakespearian lovers die for want of knowing that one of them is only feigning death, or for the jealousy set off by a dropped handkerchief, we call it tragedy, in which fate is fantastically cruel and chance is about the small but fatal detail.

In all these other, fantastical uses, chance is a synonym for outrageous luck or bad luck, a long shot or a fluke. But none goes as far as imply-ing, like Kevin, that chance is capable of breaking the chain of causality. They only say that chance means a chain we didn't expect. So the piano misses Kevin because of accident, as it turns in the air past his gorgeous black hair, owing to the way it was tipped from the balcony as the frame snagged, just when death seemed a cert. The cause was in some small detail of the angles, masses and forces, and his own antics at the lectern.

One response to life's uncertainty is to argue that there are deep invis-ible causes for everything, unlikely or not. In the words of St Augustine: 'We say that those causes that are said to be by chance are not non-exist-ent but are hidden, and we attribute them to the will of the true God.' The piano misses Kevin because God looks kindly on him. Though why, we'll never know.

---

leaves room for human choice. By linking chance and free will, Kevin is more like the American philosopher William James, who argued that chance upsets determinism to create the possibilities from which free will can choose. See also, for example, Chrysippus of Stoll, a Stoic philosopher born around 279 BC.[2]

The German poet Friedrich von Schiller said similarly: 'There is no such thing as chance; and what seem to us merest accident springs from the deepest source of destiny.' All things happen for a reason.

A similar reaction to chance is superstition, by which people hope that rituals and totems will turn fate to their advantage, perhaps by finding favour with the Gods, perhaps through mystical connection with natural power, and where today's lucky sporting mascots are the sanitised versions of Aztec human sacrifice. Fate calls the shots, but likes a bribe. The piano misses because Kevin must have slaughtered a sheep at high tide at midnight.

Even 2,000 years ago there were 'rationalists' who scorned these beliefs. Oddly, Professor Kevin would have liked their style. In Roman times, three 6s from three dice was known as 'the Venus throw'. Cicero said: 'Are we going to be so feeble-minded then as to aver that such a thing happened by the personal intervention of Venus rather than by pure luck?' He taunted astrologers with relish.

Cicero, like Kevin, defends freedom from necessity. In Cicero we see this as rationality at war with superstition. Kevin sees it as anti-rationality in favour of chaos and impulse.

As the scientific enlightenment developed in the late 17th century, earthly causes challenged mystical ones, and the extraordinary explanatory power of Newton's laws of motion led to the belief, in the physical world at least, that all matter moved like clockwork: if we knew precisely the position and motion of every atom, then in principle we could predict exactly what would happen. Thus again, the piano misses Kevin because the precise chain of causality between atoms, forces and initial conditions following given laws, plots its course, just shy of his nose. And again, there is nothing inherently random about it. Any uncertainty in this world is 'a measure of our ignorance' – as the statistician Pierre Laplace said – ignorance not of God's will this time but of the state and laws of nature; not so much 'can't know' as 'don't know'.

Two words summarise these two types of uncertainty. The first is 'aleatory', used to describe the uncertainty before a coin is flipped when we simply cannot know what will happen, often known as chance or randomness. The second is 'epistemic', used for the uncertainty when we have flipped the coin but not looked at it, in which case we don't know but in theory could – more commonly called lack of knowledge or ignorance. Although there are inevitable complications: what if it's a two-headed coin? Then what you thought was chance was tainted by ignorance.

The 'don't know' that applied to human knowledge of the mind of God and the 'don't know' that described our limited understanding of nature's clockwork implied that with more knowledge we would uncover the rigid determinism that explained human behaviour fully. Either way, everything had a cause that began outside ourselves.

This spooked people. For both kinds of ignorance – of divine cause and physical cause – seemed contrary to our internal sense of free will, and neither our own life nor other people's behaviour appears to us to obey the mechanics of a pendulum. So at this stage (and ever since), there were fierce arguments in both religion and science about whether people's sense of their own freedom of choice was an illusion.

Then came the grand age of statistics, when people were initially drawn further into the net of predictability. From the early 19th century there was an obsessive listing of deaths, crimes and, particularly, suicides. And what emerged was an extraordinary regularity, even from the turmoil of individual decisions, an order that seemed to emerge as if by a natural law and so raise doubts about the scope for chance. Because despite the choices of millions, each in unique circumstances, there seemed to be some regularising force that led to an almost constant number of suicides each year. Other patterns also emerged from the data. Francis Galton, a cousin of Charles Darwin, said famously: 'Whenever a large sample of chaotic elements are taken in hand and marshalled in the order of their magnitude, an unsuspected and most beautiful form of regularity proves to have been latent all along.' Some came to the conclusion that these predictable patterns meant that chance could not exist.

But although there was order in human behaviour en masse, there was still mystery about how this order would shape any individual. Yes, there would be a roughly predictable number of suicides, but still it was not possible to say by whom. The average, the trend and the distribution were indeed observable, but the particular was not predictable.

This led to the idea that we are all random deviations from Quetelet's average man or *'homme moyen'* (described in Chapter 13). Kevin is more than happy to be such a deviation.

There were strikingly similar developments in the physical sciences. Quetelet described his work as 'social physics', and his ideas, published between the 1820s and 1860s, may have influenced James Clerk Maxwell's development of Bernoulli's kinetic theory of gases, which describes the overall behaviour of a gas as akin to a vast blizzard of colliding individual particles. This parallel between social and physical science doesn't receive the attention it deserves, but it is a remarkable intellectual co-evolution that seems to offer a way of comparing human experience with the behaviour of matter. Here too, although it was impossible to predict the motion of each of the zillions of molecules bouncing around in a gas, even if this motion was in theory fully determined, probability could successfully explain the overall movement, just as it can describe overall patterns in people's behaviour. In fact, there is an argument that without probability we would be unable to appreciate large-scale change in people or in physics.* We would sit there watching a zillion random events among humans and molecules, with little sense of a bigger picture.

In theory, again, it was *possible* for all the molecules to wander off suddenly in one direction, as when the Improbability Drive in *The Hitchhiker's Guide to the Galaxy* was 'originally devised by physicists to transpose the underwear of a hostess at a party several feet away from her body'.[4] In practice the behaviour of the whole gas was predictable even

---

* 'If we insisted on the detailed description of the motion of individual molecules, the notions of probability which turn out to be so essential for our understanding of the irreversible character of physical events in nature would never enter. We should not have the great insight that we now do: namely, that the direction of change in the world is from the less probable to the more, from the more organised to the less, because all we would be talking about would be an incredible number of orbits and trajectories and collisions. It would be a great miracle to us that, out of equations of motion, which to every allowed motion permit a precisely opposite one, we could nevertheless emerge into a world in which there is a trend of change with time which is irreversible, unmistakable and familiar in all our physical experience' – Robert Oppenheimer.[3]

if the behaviour of the parts were in practice not. So in both the social world of people and in the physical world, the aim of predicting what would happen to each little piece, whether a human or molecule, was given up, but a new order was found in the predictability of averages.

Perhaps the best image of the behaviour of molecules is provided each Saturday evening, when 49 balls are banging around in a Perspex bucket, watched by millions of people clutching their National Lottery ticket and hoping their lives will be changed. Six balls are chosen for the jackpot – as we have seen in Chapter 12, it makes no difference which set you choose – 1, 2, 3, 4, 5, 6 or any other combination – the chance is still around 1 in 14 million that your choice will come up. Web sites keep track of how often specific numbers come up, with 38 having topped the league table of number of appearances for ages. Surely the law of averages says that 38 is due for a rest and we should keep away from it?

But this isn't how the law of averages works. What it means is that over time, even though each draw is completely unpredictable and not influenced by what has happened before, the spread of the numbers of appearances of the balls has a regular pattern, as shown in Figure 17. All these unpredictable events, when put together, turn out to have their own structure and even their own shape – the beautiful normal curve observed by Quetelet when he measured people's heights.

So although 38 has popped up 241 times – nearly half as many times again as poor old number 20's pathetic 171 appearances – this is exactly what could have been predicted back in 1994 using probability theory alone. Of course, we could not have predicted which numbers would be top and bottom, just that one number would have about 240 and one around 170 appearances.

Another wonderful illustration of the patterns of chance comes from a contender for the most boring book in the world – *A Million Random Digits*.[5] The title says it all – page after page of numbers, with no detectable order, each one utterly unrelated to those that come before. It begins with '1' and finishes with '8', an ending as unpredictable as a good Agatha Christie whodunnit. The audio version would do wonders for insomniacs. And we're still waiting for the German translation.

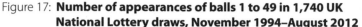

Figure 17: **Number of appearances of balls 1 to 49 in 1,740 UK National Lottery draws, November 1994–August 2012**

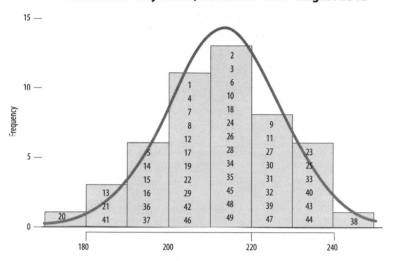

The normal curve is the predicted distribution based on probability theory.

But there are gems hidden in all this randomness. We can expect the sequence 1, 2, 3, 4, 5 to occur ten times (and in fact it occurs eleven times). And we would expect events that have a 1-in-a-million chance, such as the same digit occurring seven times in a row – to turn up once. And, satisfyingly, there is exactly one example of this – 6, 6, 6, 6, 6, 6, 6. And we can even calculate the probability that there would be no such sequence, which turns out to be 37 per cent, so we could have been unlucky.

Probability theory provides a practical way of dealing with the randomness in life – but even the patterns of pure unpredictability don't really tackle the fundamental question of whether true, absolute, irreducible chance really exists, or whether everything is at some level determined. If we knew everything about those bouncing lottery balls, could we predict which one would drop out? Sadly not, for two main reasons.

First, quantum mechanics came along in the 20th century and said that chance did indeed exist, at least at a sub-atomic level. Essential parts of the sub-atomic world could only be described as a probability. Heisenberg's Uncertainty Principle says we can't know everything about a sub-atomic particle – in terms of both where it is and where it is going. Newton's laws of causation were inadequate.

Second (and in practice more important than the quantum uncertainty), there is also an effect that throws all expectations into the air. This is chaos theory, in which even if a system were completely deterministic, in the sense that there was no randomness at any point in the chain of causation, some systems – such as the weather – might nevertheless be so dependent on, and so affected by, the tiniest difference in where they start, a difference undetectable by people, that we have no idea how they will turn out.

The standard image is of a butterfly flapping its wings and thus causing some faraway storm. A real example is Clint Dawson's experience of forest fires. He is an FBAN, a fire behaviour analyst, and his job is to try to predict the course of forest fires in Colorado. In 2012 the fires began to behave with dangerous unpredictability, often burning bigger than before. His computer models of what a fire would do became less accurate. The reason? A series of tiny changes in initial conditions, among them that no one had thought – why would they? – to factor in a change in the behaviour of beetles, of all things. A winter influx of pine beetles had desiccated the trees, making them more flammable.[6]

The *chaos* idea means that clockwork-like systems can be unpredictable simply because we can never even know exactly where they start, let alone exactly what they will do next. Causation is too complicated. Such systems might as well be ruled by chance after all (whatever that is). Had a butterfly landed on the piano just before the lads moved it, and distracted one of them, it might have made only the tiniest difference, which to Kevin would have been a difference of life and death.

Do these physical limitations on what we can know, chaos and quantum uncertainty, matter on a human scale? Is it odds-on that Kevin's trust in some fantastical element will be his doom one day? Or are these strange forces only of theoretical importance? How much faith would you have in uncertainty if a piano loomed over your head?

All risk is chance. And chance remains mysterious. Whether it originates outside people in the deep structures of matter or if it's all in our heads hardly matters to the uncertainty of living. The practical question is what to do about that: whether to strive for certainty (Prudence), to relish uncertainty (the Kevlin brothers) or to decide that life is about placing your bets (Norm).

15

TRANSPORT

NORM SHUFFLED ALONG the carriage until he found the ideal space, four empty seats, one at the window to watch the station do its strange retreat; him, his thoughts, a book on the table. He liked trains – a good bet for low MicroMorts.

A short, stout, grinning woman with bleached hair wearing an out-size white T-shirt bumped her way into the seat opposite and stuffed her rucksack under the table between her knees. Norm pulled in his feet. She settled, and he reached for his book.

Then he noticed: a wide, bright, red-and-white striped sweatband with a (badly) stitched-on smiley, looking at him, half-way up her forearm.

Norm was a man of the mind. *Four-inch sweatband*. He went through life preoccupied, tolerant, rational and relaxed ... *red-and-white striped* ... though open-minded, naturally, about personal enrichment via others' experience ... *stitched-on smiley* ... and believed each to their own, live and let live, John Stuart Mill etc. ... *half-way up!*

His eye caught the emergency hammer. He sank in his seat at the thought of hours of 'chat' with a four-inch, red-and-white sweatband and stitched-on smiley *half-way up* the kind of forearm that took its pleasures – he didn't doubt – tearing off testicles.

Thinking fast, he took a firm grip on his emotions and began to calculate the risks. 1: Assume successful avoidance of eye contact; 2: Combine low-to-moderate chat hazard with 3: significant threat of, if not spontaneous psycho violence then at least embarrassment-grade nutterness; plus 4: (he winced, but there it was) probable class and diction issues – Glaswegian? – and arrived at a 51 per cent probability of hell. Approximately.

She reached into her rucksack and pulled out ... what?

52 per cent, thought Norm.

A large, red, paper napkin that she unfolded onto the table ...

54 per cent.

... spread well onto Norm's side ...

58 per cent.

reached down again, paused, and placed one-by-one on the napkin: an all-day-breakfast mixed-triple-pack sandwich of egg mayo, spicy sausage, and bacon and tomato ...

68 per cent.

a small pork pie ...

75 per cent.

crisps, a Ripple, Crunchie and Mint Aero ...

82 per cent.

and a can of Strongbow.

95 per cent.

Norm whimpered.

She rotated the Ripple 180 degrees.

99 per cent. Holy shit.

Norm was a dead man. Paralysed. Waiting for the cobra to strike. Surveillance blown, he stared floodlights. The only option was to get off at Crewe. Was ever a woman of more menace? Every move ended with a pause – and another smile. Every smile ratcheted Norm's unblinking terror.

She picked up the can, stared at the label, smiled and pulled the ring. It hissed.

She paused.

She tore it back – and set the can carefully beside the Crunchie with her right hand – *sweatband half-way up*. She paused. *Stitched-on smiley*. She lifted the can and drank, deep and long. He watched her throat's gristly dance. She put down the can, paused and smiled.

Please, he thought regarding the pork pie, let her not spit. Then a more awful thought crossed his mind: what if she offered ...

She caught his gaze and tilted the can with a quizzical eyebrow. Oh God.

Please don't say *swig*, thought Norm. 'Erm, thank you, thanks. I won't actually. Just had breakfast, big one, already, really, coffee, tea ... orange juice, coffee, honestly. Already. Thanks.'

She nodded, and turned to face the man who had sat next to Norm, hidden but for his fingertips, behind the *Financial Times*.

'If you don't mind me saying,' she said in cut-glass English, 'it's probably a weakness of demand, at least in part, don't you think? A dab of Keynesian stimulus wouldn't be out of order. The economy, young man. Well, you can tell by the data on corporate investment that it's as flat on its back as a tart.'*

The *Financial Times* snapped straight.

She smiled and looked down at the table-top, placed her palms either side of her selection, studied it, picked the Crunchie, unwrapped it with delicacy, raised it to one nostril and inhaled. She moved the Crunchie towards the side of her mouth, closed her eyes and then, as if amplified, pulverised the end between her molars.

She sat back, opened her eyes and crunched, like a laird pacing the long gravel drive of his estate, and smiled again into the middle distance.

WHAT'S HAPPENED TO NORM? The rational paragon just became a mouse. There he was, nicely archetypal, come over all paranoid because of a wristband, an accent and a pork pie in his personal space. Has he lost the plot? Well, yes. But then, we do. People are as inconsistent about danger as other things – often for good reason.

They may, for example, lose their nerve. John Sergeant was a BBC correspondent who covered conflicts including Vietnam and Northern Ireland. Then he was taken hostage in Cyprus and held at gunpoint for 33 hours. After that, he said, he didn't want to do it any more. He

---

* About which she was at least half-right, though whether Keynesian stimulus was the solution is not for us.

realised that he was 'extremely frightened' and went into reporting politics.

Something similar happened to David Shukman. He'd reported from war zones for 15 years. One day, asked to board a clapped-out helicopter from Tadjikistan to Afghanistan shortly after 9/11 – he refused. And that was that. The more he reflected on his assignments, the more nervous he felt. He also thought more about the price paid by his family.*

Stories allow people freedom to change. Probabilities could be taken to suggest that people should stay the same as long as the numbers stay the same. And one response to people who change their minds is to say that it proves we can't judge danger sensibly. For it's not as if the outside world has changed, as if train travel for Norm is more dangerous now, or war reporting – or war – wasn't dangerous enough before. In terms of probabilities, the risks to Norm are the same before and after the pork pie. All the change is on the inside.

There are two simple justifications. The first is new information (see also Chapter 14, on chance, about epistemic uncertainty). Norm learned something about the risks of train travel that he hadn't known or experienced before. Doctors, governments, the Health and Safety Executive also all revise their advice about risk as they learn. But true risk – whatever that might be – is unknowable, since our understanding can always be bettered. Always lay your sleeping baby face down, they said with confidence, until the evidence changed 180 degrees. 'If the facts change, I change my mind', said the economist John Maynard Keynes, and the facts about risk are seldom final.

The second justification for changing your mind is that you see danger in a new light, like David Shukman when he thought more about the pain his family felt, or as Norm does when his prejudices are stirred.

So to expect lifelong consistency from Norm would be inconsistent: it's not normal. Norm's basic approach to life – that reason and

---

* One of the best-known literary losses of nerve also comes in an account of war. Yossarian in Joseph Heller's *Catch-22* sees an exceptionally gruesome death and decides that the enemy is anyone trying to kill him, and this includes his own side. The world becomes more deadly even as it remains the same.

calculation are the one true way – will survive, you'll be glad to know, but perhaps with more sense of his own ups, downs and weaknesses. All the same, he has taken a knock.

But he would still want facts about the quantifiable risks of travelling, and here they are, first by train. Roads and air follow.

The hazards began – badly – on 15 September 1830, with the opening of the Liverpool to Manchester railway. William Huskisson, a leading English politician, went to greet the Duke of Wellington at his carriage and was flattened by George Stephenson's *Rocket* coming up the other track. His leg was 'horribly mangled', and he died a few hours later. Oddly, his death attracted huge and welcome publicity to the railway. This lesson – that it is safer to be inside a train looking out than outside looking in – is still valuable.

With qualifications: the inside of a train is also a public space. Of itself, the presence of other people – wristbands included – has only the tiniest effect on your chance of death or injury. Norm's personal space invasion makes him quake, but with no basis in mortality statistics – murders on trains are a good plot device but extremely rare, although a man was jailed for life in 2006 for the murder of a student on a Virgin Express from Carlisle to Devon who did no more than catch the man's eye: 'What the f... are you looking at? I'll stab you in a minute', which he did, with an 11cm kitchen knife.[1]

For lesser violence, the risk is also real and also small. A stranger on a train might thump or abuse you, and 3,300 assaults were recorded on trains in the UK in 2010. Even so, this is 1 in 400,000 journeys, and so if you travelled every day on a train you might expect to be assaulted once around every 1,000 years, if the risk were equally shared.

Still, this is little comfort to Norm, who might fear that this is his thousand-year reckoning. Or maybe prejudice distorts his sense of the odds. There are no statistics for 'class and diction issues', as Norm puts it, or other awkwardness like bad language, noise, being pestered or leered at, or the smell of someone's burger. But these things happen. So is it this – unpleasantness rather than physical injury – that people are really worried about when they say they feel unsafe on public transport? Is background irritation part of what makes the company of strangers menacing, even when the risk of violence is remote? Norm is afraid because he can't predict wristband's behaviour – 'pyscho violence' or 'embarrassment-grade nutterness?' – and we feel more wary of strangers than friends or relatives on the back seat, depending on who your friends and relatives are. As one research paper put it: 'feelings of anxiety and psychological factors act to make some people feel uncomfortable on public transport and [...] this acts to increase perceptions of poor personal safety.'[2]

Other influential factors were gender – women tend to be more anxious than men – and 'the actual experience of a personal safety incident'. Mind you, even the most pronounced of these influences was found only to tinker at the edges of what makes most difference of all to how we feel, and that is the accident record itself. So the numbers do matter.

So, more numbers. Despite William Huskisson's misfortune, train travel grew massively in the subsequent 180 years. In Great Britain in 2010 there were 1,400 million journeys, up from 800 million 30 years ago: that's 4 million journeys a day, totalling 54 billion kilometres over the year.[3] This means that each one of 60 million people does on average 900 km (560 miles) a year, or around 10 miles a week, although this is one of those wonderfully misleading 'averages' that reflects the experience of nobody. Who travels 10 miles a week by train? There is huge variability, ranging from the dedicated commuter to someone in the countryside, cut off from trains by the Beeching cuts in the 1960s, to whom a train trip is a rare treat (or not).

At the bottom of this distribution will be sufferers from 'siderodromophobia', the inordinate fear of trains, but most people feel a reassuring solidity and familiarity about train travel, as does Norm, usually. And with good reason – it is extraordinarily safe to be a passenger on a British train. There were no fatalities in rail accidents for on-board passengers in 2010, nor in the preceding three years. Eight passengers were killed in stations: an elderly man falling down an escalator, four people falling off platforms when intoxicated, and so on. This is a rate of one death per 170 million passenger journeys, and even then does not reflect the experience of the typical passenger.

Just because there have recently been zero deaths in rail accidents does not mean the future risk is zero, and we can fit a smooth trend to past data,[4] which suggests the risk has been dropping at around 6 per cent per year and is now expected to be 1.6 fatalities per year. This corresponds to 33,750 km (20,000 miles) per MicroMort. The Rail Safety and Standards Board uses a slightly different method and claims around 7,500 miles per MicroMort, around 30 times safer than a car per mile.

People do fall over on platforms, but the 240 major injuries recorded in 2010 work out at less than one for each 5 million journeys: the accident rate is higher at off-peak times, when users may be less familiar with the system, or drunk, or both. Maybe that's their fault. Though a guard on a train was jailed for five years in 2012 after signalling for the train to leave while a young girl leaned drunkenly against it. She fell between the carriage and the platform and was crushed.[5]

How does the UK compare with other countries? From 2004 to 2009 it had the lowest fatality rates in the EU except for Sweden and Luxembourg (which has only 170 miles of rail).[6] In contrast, 200 people were killed in just three accidents in 2010 on Indian Railways, although they do carry over 30 million passengers a day. But the UK has had its own share of spectacular disasters on the way to safer train travel, such as a triple collision at Harrow and Wealdstone in 1952 which led to 112 deaths. After the Second World War the number of passenger deaths regularly exceeded 50 per year, while around 200 railway workers were also killed each year: in 2010 the number of workers killed was 1.

But as Huskisson found out, a moving train can be lethal if you are not in it: 239 people were killed by, rather than on, British trains in 2010. These were mainly suicides and suspected suicides, but 31 were crossing the line at a legal or illegal place, about half the usual figure. The number of non-passengers killed has not improved: it was almost exactly the same in 1952, when 245 died. The number of suicides has also been amazingly constant, varying between 189 and 233 in the last ten years. Quetelet would have understood.

In Victorian days a train crash with a few fatalities might only have made the inside pages of a newspaper. Now the image of a derailed train would have massive coverage. It is clear that just counting the bodies fails to measure public interest and concern at a 'disaster'. One accident that kills, say, 10 people gets far more attention than 10 accidents that kill 1 each. The 250 people each year who die on the track usually have little publicity, but imagine if they all occurred at once ...

That's a problem if you're deciding whether to spend money on safety, where the government uses the concept of the Value of a Statistical Life (VOSL) (or Value for Preventing a Fatality – VPF). As described in Chapter 1, this currently stands at around £1.6 million, corresponding to £1.60 to avoid a MicroMort. But the calculation falls apart if lives lost in a group are somehow worth more than lives lost individually. So a multiplier can be used that 'up-weights' lives lost in a 'disaster' to reflect public concern.

Although railways are generally seen as safe, people still seem to think it a good idea to spend huge amounts of money making them safer. The Rail Safety and Standards Board employed a consultant philosopher to help them understand why. He concluded that it was a moral issue – people quite reasonably had a sense of shame that train disasters could happen, felt outraged that they were caused by someone's actions and so didn't tolerate much risk.[7]

But caution can have unintended consequences. After the 9/11 attacks in New York many people felt more nervous of flying and took to their cars instead. Gerd Gigerenzer states that 1,500 more people than usual were killed on US roads over the following year.[8] Similarly, after a train derailment at Hatfield killed 4 people in 2000, speed restrictions on the rail network while track was checked, caused the system to clog up, again people took to their cars, and it has been claimed that an extra 5 or so were killed in the first month (although it is unclear where this figure comes from). Better to stick to the train, as long as you avoid alcohol and escalators – and don't mind pork pies or accents.

Roads

Roads are unquestionably more dangerous. In 2010 our average annual risk of dying on the roads was about 31 MicroMorts. Since the average dose of acute fatal risk from external causes is about 350 MicroMorts a year (roughly one a day, remember) this means that something like 9 per cent of that risk in the UK is on the road. Relative to trains and planes, it is on average far more dangerous, per mile, to drive, ride a motor bike, cycle or walk wherever there is traffic.

But how many believe they're average? Most think they're better than that – a self-confidence known, naturally enough, as the above-average effect or illusory superiority.[9] Like everyone else, even Norm once thought he was better than average. This is illusory for obvious reasons, since only half of drivers can be in the top half, so if exactly half of drivers thought they were in the top half they could in theory *all* be wrong. Even if you're genuinely better, you might not be much better. And if above-average driving ability turns you into an above-average jerk, you might be so cocky that you become a bigger hazard.

Illusory superiority is also linked to an illusion of control: the idea that with the wheel (or handlebars) in our grasp, our fate is in our own capable hands, not at the mercy of the pilot, who loses it and takes the plane down over a poisonous divorce. But although sitting in economy class, trapped, waiting for the wings to fall off certainly *feels* vulnerable, lack of control evidently does not equal high risk. MB had no control whatsoever during a simple heart operation. He didn't say: 'Let me do that.' Even so, to be busy at the wheel with silken mastery of road and vehicle, fully in control (you think), can give false reassurance and feed self-belief. For risk on the road is clearly determined not only by our own ability but also by that of all the other drunks and lunatics out there, and partly by accidents which have nothing to do with anybody's skill but are out-of-the-blue hazards like the stray poodle that runs out.

So again the question: are we mad? One tactic for overcoming fear of flying is to imagine you are at the controls, pulling back the joystick as you take off, like a child pretending. Clearly, we are fooling ourselves. And we know it. We are not that stupid. Perhaps what the pretence does is to remind us that at least somebody is in control, and this is what they will be doing, and we approve.

There is an argument that this double standard – the risk when I'm driving is OK, the risk on a train or plane isn't – is justified even though it seems to reverse the statistical evidence. That argument is backed up by the Rail Safety and Standards Board's moral philosopher: if others are in control, of course they ought to take more care of me than I might take care of myself. That's their job. If I risk my own neck, it's reckless; if *they* risk my neck, it's criminal. Hence, runs the argument, there's logic in the 'pages 1–12' press coverage of a rail death – someone else is to blame, and the accident is also about public trust, corporate and government responsibility – compared with the 'news in brief' treatment for death on the roads, which are more like fatal DIY, a private misjudgement after which private grief is of little public importance.

You can argue with this, but while we can mock people who prefer the safety of cars to flying, some of this preference is better seen as an expression of trust, or mistrust, than proof of irrationality.

A final complication is that one response to feeling safe is to drive more dangerously. This is known as 'risk homeostasis', the idea that we have an inbuilt risk thermostat, a level of risk we're prepared to tolerate (or a level of risk that we seek out and try to maintain because that's the level we enjoy). As the risks change – with seat belts, air bags, better brakes – we adapt our behaviour to maintain the same level of risk. So, as cars are made safer, many drive faster, and transfer risk to other road users, such as pedestrians, who don't have air bags.

According to John Adams, a specialist in transport risk, safety would be best achieved with a huge spike on the steering column pointed at the driver's chest and no seat belt. That should recalibrate the risk thermostat of a boy racer or two to about 10 m.p.h. Dudley Moore said the best safety device was a cop in the rear-view mirror. Both in their own way argue that people become more careful the more that it hurts not to be. If you want them to drive safely, expose them to danger. The theory of risk homeostasis also works the other way: if we do protect people from the consequences of taking risks, they will take more of them. How this balance of mechanics and psychology pans out is hard to tell in advance. Will the new safety device make more difference to the casualty figures than any change in behaviour?

Adams also argues that one reason why road casualties fall is that some roads get so lethal that no pedestrian will go near them. If true, falling mortality on these roads is a measure not of safety but of danger.

In short, the risk of death or injury on the roads is subject to all manner of psychological filters and questions of interpretation. Are the data useless, then? In fact, what they seem to say is so stark, the trends so clear, that they can stand a big margin of error or interpretation and still be clear-cut. All data are wrong. The question is whether they're so wrong that you can't draw conclusions from them. Here they are.

In 1950, a few years before DS was born, there were 4.4 million vehicles registered in Britain. In 2010 there were eight times as many, which after the growth of the population equals a rise from one vehicle for every 11 people to one for every 2 people.

A reasonable assumption might be that more vehicles on the roads means more deaths. Statistics suggest not. There were 5,012 fatal accidents in 1950. By 2010 this had dropped to 1,850, a fall of 63 per cent in absolute terms.

In relative terms the fall is even more extraordinary, given that there was a rapidly rising volume of traffic. For every 100,000 vehicles in 1950 there were 114 deaths. In 2010 there were 5, a 96 per cent reduction. When DS remembers how, as a child, he enjoyed riding in the front seat of their old van, which had no regular safety inspection, seat belts or air bags, and then how people used to drive to the pub, drink all evening and meander home, he is not surprised. It is a fall from an average in 1950 of 102 MMs per person per year to about 31 MMs per person per year in 2010.

Among car occupants, the number of fatalities are much as they were 60 years ago – the figure was around 20 a week in 1950, rose to 60 a week in the 1960s and now is back at 20 a week. The main saving of life has been pedestrians and cyclists: from 60 a week in 1950 (although in 1940 it was even worse, when a staggering 120 a week were killed by unlit vehicles in the black-out), down by over 82 per cent in 2010 to 11 a week. Another way of looking it this figure is that, for every 100,000 vehicles, about 1 pedestrian or cyclist is killed per year. (We will compare this with other countries later.)

These statistics depend on counting bodies, which is grim but easy. Counting accidents and injuries is trickier – what is an injury anyway, and how bad does it have to be to be recorded? – but it's reported that in 1950 there were 167,000 accidents, with 196,000 injuries, and in 2010 almost the same: 154,000 accidents, with 207,000 injuries.

This means that people still crash into each other about 400 times a day, but the proportion of these accidents that are fatal has dropped staggeringly, because of speed limits, safety features and improved and quicker medical care.

Almost all the richer nations have seen this trend: in the 30 years from 1980 to 2009 road fatalities fell by 55 per cent in Australia, 69 per cent in France, 63 per cent in Britain, 54 per cent in Italy and 58 per cent in Spain, in spite of increasing volumes of traffic, although the reduction in the US was only 34 per cent, and deaths actually rose slightly in Greece. For countries that collect the relevant data, we can work out an average number of MicroMorts per 1,000 kilometres a vehicle travels: 4 MM per 1,000 km in the UK, 7 MM in the USA, 10 MM in Belgium, 20 MM in Korea, 40 MM in Romania and 56 MM in Brazil.[10]

But who bears this risk? In richer countries the majority of road fatalities are the occupants of cars, whereas in poorer countries it's what are known as 'vulnerable road users': pedestrians, cyclists and those whole families squeezed onto a single small motor bike. In Thailand 70 per cent of all road deaths are riders of two-wheeled vehicles,[11] which will not surprise anyone who has witnessed the traffic in Bangkok.

And it's in low- and middle-income countries that people are exposed to the serious risk. It's estimated that around 1.4 million people are killed on the roads each year, which is around 3,500 a day, and of these 3,000 are in the developing world – in spite of those countries containing less than half of all cars.[12] The majority of these victims will be vulnerable road users. South Africa sees 15,000 killed each year, a statistic brought into sharp relief when Nelson Mandela's great-granddaughter Zenani was killed.

The World Health Organisation (WHO) predicts that road traffic injuries will go from its current 9th position in the cause-of-death league table to 5th place by 2030, causing 2.4 million deaths (as well as between 20 and 50 million injuries), largely of young people, and at enormous cost to the economies.[13] The WHO points out that only 47 per cent of countries have laws about safety features such as speeding, drinking, seat belts, helmets and child restraints. And these are often not enforced.

The average risk for an individual from dying on the roads – 31 Micro-Morts per year in the UK – is 103 MicroMorts in high-income countries generally, but 205 in low and middle-income countries. It may seem odd that countries with fewer cars are riskier, but, perhaps surprisingly, deaths per vehicle decline strongly as roads get busier. It is empty roads that are really lethal. This observation even has a name: it is 'Smeed's law' of traffic.[14] Smeed's law is reflected in some extraordinary statistics about Ethiopia, which the WHO report as having only 244,000 registered vehicles in 2007, but which nevertheless managed to kill 2,517 people. The majority of these deaths were pedestrians – that's 1 death per year for every 100 vehicles: the same rate applied to the UK, with 34 million vehicles, would mean 340,000 people killed each year, instead of under 2,000.

One lesson is that – if we have the money – the bigger a danger becomes through faster and heavier traffic, the more likely are we to do something to control the risks. Risk doesn't just sit there as a settled fact of life, and it can't be separated from how we react to it.

Flying

'The pilot has advised that there is some turbulence ahead. Please return to your seats and fasten your belts.'

As the plane bucks around, the wings flap up and down, babies cry and the knuckles turn white, is there anyone who remains completely calm, except those who are drunk, drugged, asleep or all three?

Fear of flying, or aerophobia, is common. Around 3–5 per cent just won't fly, around 17 per cent admit to being 'afraid of flying' and 30–40 per cent have moderate anxiety.[15] We have known risk professionals – archetypes of rationality – who refuse to fly.

It is a treatable condition, and British Airways runs day courses for around £250 that conclude with a 45-minute flight.[16] Sadly they do not publish their success rates, or the counts of those carried gibbering from the plane.

Again, of all the classic fear factors, it is the feeling of complete lack of control over our fate that kicks in hardest when strapped in a plane with the ground a long way off. Perhaps a lack of understanding also contributes – how does it stay up anyway? If we all stopped believing, would it fall from the sky?

And we've seen how the easy availability of negative images influences

Figure 18: **How the flight-time splits, and when accidents occur, 1959–2008**[18]

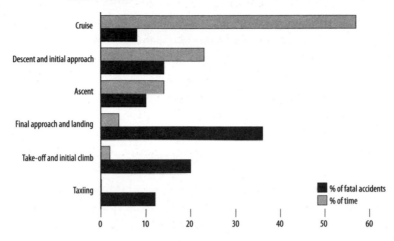

our perception, and the media certainly like to linger on pictures of wrecked planes. Those of a certain age bring to mind examples from history, from Buddy Holly to the Manchester United football team.

But for real aero-disaster porn nothing can beat the unsubtly named Plane Crash Info website,[17] which keeps a running database of all fatal crashes involving commercial airlines, together with lurid photographs and even audio clips from the black box recorder – which are disturbing, to put it mildly. In 2011 it recorded 44 crashes, around 1 a week. Sounds bad, but it's an improvement on 70 in 2001 (including, of course, 4 on 9/11).

Their analysis, vividly illustrated in Figure 18, shows that the cruising part of the flight is by far the safest, taking most time and with fewest accidents. Per minute, take-off and landing are around 60 times as dangerous as the middle of the flight. Try muttering this mantra during turbulence.

Plane Crash Info also estimates that human error accounts for over half the accidents – though whether it's pilot error or mechanical failure, there is still nothing you can do about it.

But perhaps you can do something about which airline you fly on: Plane Crash Info estimates that, in the safest 30 airlines, there is a fatal

Figure 19: **Causes of fatal accidents on commercial airlines, %, 1950–2010**[19]

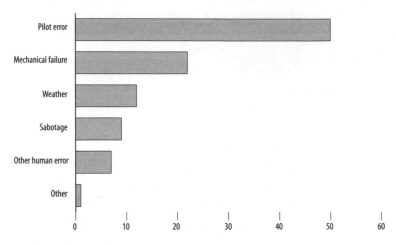

accident on average every 11 million flights, and since there is some chance of surviving, the odds of being killed are around 1 in 29 million per flight. In contrast, for the bottom 25 airlines, there is around 10 times the rate of fatal accidents, and your chance of being killed on a flight is about 20 times higher.

For calculating MicroMorts, the best data come from the US National Transportation Safety Board statistics on planes carrying at least 10 passengers.[20] Looking at the ten years from 2002 to 2011, each year US commercial airlines flew an average of 10 million flights. Over that period there were no major disasters, and an average of 16 passengers and crew were killed each year of the 700 million who got on a plane – this works out as 0.02 MicroMorts per flight, or an average of 50 million flights before you are killed. So if you took one flight a day, that's 120,000 years.

Now, you may complain that we have neatly avoided 9/11, in 2001. So if we look at 1992–2011, during which there were a number of major accidents, it works out at 0.11 MicroMorts per flight, or an average of 9 million flights before you are killed.

But how should we measure these risks so that we can compare them with different ways of travelling? By journey, by mile or by hour? Let's use as a benchmark the somewhat pessimistic figure of a 1-in-10-million chance of being killed per flight, which is 10 flights per MicroMort. An average US commercial flight lasts 1.8 hours and travels 750 miles. So that works out as 7,500 miles or 18 hours per MicroMort, around 20 times the distance for driving in the UK, and similar to rail.

As we've seen, risk is not evenly spread through the flight. And if you were really going to choose transport for a trip on this basis, you would need to factor in the drive to the airport, or the cycle to the railway station and so on.

But one set of flying statistics stands out from the rest and these are a different matter: small private aircraft, known as 'General Aviation'. There are around 220,000 registered general aviation aircraft in the US, and each year around 1,600 have accidents, about 300 of which are fatal. That's 6 a week. On average over the last ten years, no fewer than 520 people a year were killed in small aircraft and helicopters, 97 per cent of all air fatalities.

This works out as 13 fatal accidents per 1 million hours flying, around 150 times the rate of commercial airlines. That's around 1 MM for every 6 minutes in a small plane, travelling perhaps around 15 miles, about the same risk, per mile, as walking or cycling.

We see the same pattern in the UK: between 2000 and 2010 there were 9 fatalities on airlines, 34 in helicopters and 202 in 'general aviation'.[21] Bigger is definitely safer. So here's a question on the front line of stats versus psychology: do pilots of small planes feel safer for being at the controls? If you know any, ask.

One last thought. Turbulence can be dangerous even if the wings stay on: 13 people were injured during turbulence in the year up to August 2012 on US commercial airlines. Of these, 12 were cabin crew.[22] So when the seat-belt light goes on, buckle yourself in and just be pleased you are a passenger and not pushing the drinks trolley.

16

EXTREME SPORTS

I T WAS A SLOW DRIVE up eleven hairpin bends on the road known as the Troll's Ladder, then a slower climb on foot following the cairns, avoiding loose rock up to the giant, conical mountain top called Bispen, 'the Bishop', standing like an upturned ice cream cone.

Here the air was still, clear, cold. Far below, far away, the lake was a hard spangle of light between hills and granite. High, thin clouds watched over the valley. Here the world of people disappeared. Here was nothing but the view and the drop, landscape, distant shapes and textures, shades of grey and green.

On the edge stood Kelvin, alone and scared, as scared as anyone else on the edge of a mountain. Then he stepped off.

From above, he was a wrinkle plunging to the ground. From below, he was a stick man etched against the blue. But in his own space, Kelvin had stretched out the webbed arms and legs of his flying suit against a riot of noise – and was free.

'You just kind of go wherever you look,' they said about base-jumping from mountains in a wing-suit, 'just kind of ... stick your arms out.'

At near-terminal velocity he planed out, swooped into a glide. On previous jumps he had 'cleared out the space', as they put it, got the hell out of there and flown away from the mountainside. But that got boring. He tilted his shoulders to veer towards the cliff, measuring his line past the vicious hard road of the rock face over flashing shrubs and bare pronged trees, past jots of reflected light and sudden shadows as the land rose and fell in spikes and bumps and hollows and crevices, and beyond them all in roaring buffeting air he curved and scored the long, smooth line down.*

Buzzing the walls, they called it, hands outstretched, a fingernail from pieces. 'Mess up and you will 100 per cent die,' they said. 'Click. Like that.' Kelvin knew it: knew it, that is, as well as any of us can know that our own death is real.

'Sick, man, real sick,' they'd be saying in a, like, totally envy-racked and mind-blown way as Kelvin bombed past the watching crowd on a Trollsteigen hairpin at more than 100 m.p.h., here and gone, a whistling projectile so vicious that he could smash a hole in your house but was, by the way, a human being. Cool.

For another minute he roared above forest, a flying squirrel embracing the air. Not really flying, 'more like falling with style,' they said. 'The most extreme badass activity on the planet,' they said.

And as the lake drew closer, Kelvin pulled in his arms, brought his heels together and stood up, still blazing over the water, but slowing now he pulled his cord and released the chute. And then, violently, there was peace. Only the gentle sound of Kelvin's laughter and bottomless exhilaration, the topless high as he steered towards the shore.

And as he drifted down, he thought of those who said this was impossible. Well, guess what? He just did it. He appreciated people who thought he was crazy. He thought of Graham Greene playing Russian roulette with his father's revolver, and some memory he scarcely recognised as his own about Dostoevsky and a piano.

'Sick! Kelvin man!' They said, as he walked back with his chute.

'Dude!'

'Sick!'

'Yo!'

'OK!'

No doubt about it: near death was real living.

---

* For the most jaw-dropping, eye-popping, vicarious risk-taking, watching wing-suit flying is hard to beat. Try this video.[1]

THERE IS A BIANNUAL competition for jumping off cliffs in a wing suit. It is, incredibly, a race. The aim is to be fastest down. It is called the Base Race and is organised by the perfectly named (for a risk-taker) Paul Fortun from Germany. You'd think, unlike Kelvin, he'd be fearless. How else could he step off a mountain? But in 2012 he said: 'I'm terrified every time I jump, everyone is. I think if you're not scared of something like this, then why bother?'[2]

Which is a neat inversion of the precautionary principle – do it *because* it might go wrong. 'It's all about the taste of fear and lack of control,' he said. 'I love it.'

Much of this book discusses mortal danger as if it's a bad thing. But danger can be a scream. Oddly, those who love it and those who hate the idea needn't necessarily disagree about the quantified risks. Paul Fortun and Kelvin do it not because their perception of the probability of death is different from other people's, but because it's much the same. That's why they like it.*

Even when life was brutish and short, with daily risk from disease, hunger and conflict, and you might guess that not many would choose to stand, or jump, sportingly in harm's way – there being quite enough harm around already – people boxed bare-knuckle from the early 16th century, played or more accurately fought in the kind of mass football slugfest without referee or rules that occasionally left people dead and maimed, or, if rich, they jousted (in medieval times) and (later) galloped full-tilt over hedges.

As life became more cosy and predictable for some of the richer citizens of 19th-century Europe, there was the Alpine Club – membership criterion: 'a reasonable number of respectable peaks' – founded in London in 1857, when the British dominated mountaineering and bagged the classics in desperately unsuitable clothing. These gentlemen – and one or two women – had many motivations. For some it was science, for others a spiritual union with nature, for a few 'the cool keen finger of danger'.[3]

---

* Although this is unusual. On the whole, if people like the benefits of risk-taking, they tend to think the risk is objectively lower, even if still high enough to be thrilling. See index for 'affect heuristic'.

Climbing mountains was risky then and can still be risky now. Apart from the possibility of falls, there is low oxygen, freezing temperatures, exposure to wind and sun, and exertion. But it's difficult to put numbers on these risks. For although it is relatively straightforward to count dead bodies, do we consider these deaths as a proportion of all who go climbing, or of only those who attempt high peaks per day of activity? Another problem is that, whatever the measure, we need to know how many people get up to whatever it is that turns them on, and these data can be hard to find. So the following numbers are bound to be rough and ready.

By the end of 2011, 219 people were known to have died climbing Mount Everest, 1 for every 25 who reached the summit. Of the 20,000 mountaineers thought to be climbing above 8,000 metres in the Himalaya between 1990 and 2006, it was estimated that 238 died, a rate of around 12,000 MicroMorts per climb.[4] In another study of 533 mountaineers on British expeditions above 7,000 metres between 1968 and 1987, there were 23 deaths (1 in 23), equal to about 43,000 MicroMorts per climb.[5] Mountaineering at these levels is riskier than a bombing mission in the Second World War, or about equal to 117 years of average acute risk.

Next, falling with style. For while some are terrified at the thought of stepping into an aeroplane, others enjoy stepping out mid-flight, or otherwise hurling themselves into the air. The dangers have been apparent since Icarus first strapped on fake wings and discovered that the birds had made it look too easy.

Parachutes were developed in the late 1700s, hang-gliders a hundred years later. This was a time when inventors were expected to demonstrate their bolder ideas personally. One such was Franz Reichelt, in 1912. He was a tailor who invented a wearable parachute, like a cross between a voluminous raincoat and an inflatable dinghy. It worked with dummies.

He assured the authorities that it was a dummy he wanted to test at the Eiffel Tower. At the last moment he clambered on top of the railings wearing the parachute and could not be dissuaded. Early film shows the heart-stopping moments as he rocks back and forth, finding the courage to jump, then flutters forlornly down like a shot bird. The crowd, silent in the film but doubtless screaming, rush forward to see his body on the frozen gravel.[6] The police are then seen measuring the depth of the dents.

Parachuting competitions began in the 1930s, and the US Parachute Association estimated that an average of 2.6 million jumps were made each year between 2000 and 2010.[7] But it's still not entirely without risk: in that period 279 people died, around 25 a year for a rate of about 10 MicroMorts per jump. Examination of the details of the accidents reveals these were mainly experienced enthusiasts with thousands of jumps behind them, pushing things a bit far. Novices were safer.

Reichelt could be considered one of the first base-jumpers, a sport in which sky-divers leap from fixed objects rather than planes. This, it hardly needs saying, is dangerous, although, perhaps surprisingly, it can be less dangerous to jump from certain mountains than to climb an especially tall one. The Kjerag Massif in Norway is one of the safer launch points, with a sheer 1,000-metre drop. In theory, this allows ample time to do something to soften that inevitable reunion with the ground. Nevertheless, over 11 years and 20,850 jumps, there were 9 deaths and 82 non-fatal accidents[8]: that's one death in every 2,300 jumps, or an average of around 430 MicroMorts per jump. And this is one of the safer spots: base-jumping is not a mass-participation sport, for obvious reasons, but 180 deaths have been recorded, increasingly featuring wing-suits that open up like a flying squirrel, much as poor Reichelt originally intended.

If diving through the air doesn't appeal, there's always diving underwater. Jacques Cousteau's development of the 'Aqua-Lung' in 1943 turned scuba into a leisure activity, and the British Sub-Aqua Club (BSAC) now has more than 35,000 members. The BSAC keeps a careful tally of diving fatalities and recorded 197 deaths in the 12 years from 1998 to 2009, an average of around 16 a year.[9] It estimated about 30 million dives over this period, so the average lethal risk was around 8 MicroMorts per dive. But this is an average – for BSAC members it was 5 per dive, for non-members 10.

Diving, like modern mountaineering and parachuting, relies on technical developments to reduce the risk of an inherently dangerous pastime. By comparison, running seems natural and moderate, one for the more cautious. But long-distance running, as Pheidippides found 2,500 years ago, after collapsing and dying from his run of 26 miles to announce victory at the Battle of Marathon, can end in more than sore feet. In 3.3 million marathon attempts in the US between 1975 and 2004 there were 26 sudden deaths.[10] That's around 7 MicroMorts a marathon on average, roughly like a scuba dive or a sky-dive. Recent attention has focused on the risks of drinking too much water, from which one participant in the London Marathon died in 2007.[11]

Most weekend sports aren't likely to kill you, but injuries are common.

Data used to be collected on a sample of people treated in hospital for accidents from leisure activities, and the resulting Leisure Accident Surveillance System (LASS) statistics record, for example, 620 people injured in riding schools in 2002.[12]

But this was based on only around 17 of the hospital accident departments in the UK. Scaled up to the whole country, the estimate is of about 12,700 accidents at riding schools. Similarly LASS estimates 6,500 accidents at golf clubs, and nearly 700,000 injuries from all sports, of which the majority – 450,000 – were from ball games, enough to make you want to stay home and watch the football on television. LASS even identifies the specific object associated with each injury: 200 from javelins (best not to ask), 1,600 from skipping ropes, 17,000 from cricket balls, 260,000 from footballs, 34,000 from skateboards and rollerblades, 3,200 from fish hooks and 5,800 from bouncy castles. And 164 cricket stump injuries.

MB asked a budding *traceur* – a practitioner of parkour, or freerunning – about the injury risks of rolling, jumping and vaulting up, down and over urban obstacles at speed, and was ticked off for not knowing that parkour was a technique for avoiding risk by learning to move swiftly but safely. To which you can't help replying, if safety is the point, why not take the stairs? (Though stairs have their hazards too: see Chapter 18, on health and safety.)

Clearly there's more to it than being careful. Equally, most of them don't want to die. Taking sky-diving, scuba-diving, marathon-running and other moderately extreme sports, there seems to be some natural level of risk – up to around 10 MicroMorts – that most participants are prepared to accept while remaining reasonably sensible. This does not include base-jumping or climbing to high altitudes.

So it is not illogical for people to talk of safety even as they flirt with danger. Dangerous sports do not equal recklessness, for the risk seems to be carefully, if not always consciously, calibrated. The acute risks of these activities are certainly higher than those of an ordinary day at the desk, but for most participants of most extreme-ish sports, the risk of ten times the average daily dose seems to be about where their risk thermostat is set at weekends.

One study by Stephen Lyng of what he calls 'edgework', a phrase coined by Hunter S. Thompson in *Fear and Loathing in Las Vegas* (see also Chapter 9, on illegal drug-taking), describes edgework in the bracing language of social theory as 'acquiring and using finely honed skills and experiencing intense sensations of self-determination and control, thus providing an escape from the structural conditions supporting alienation and oversocialization'.[13]

Hate the job, feel like a cog in the machine, tired of the daily Micro-Mort? Want to feel spontaneous and forget your everyday self? Then squeeze about 10 days normal risk into one – and jump.

As with drugs, risk perception is capable of turning every threat into an attraction. But it usually stops short of making death desirable. Some speak of a lack of control, others of the fine judgement that enables them, just, to keep control. Either way, they think they will survive to tell the tale, and they always have the option of leaving the danger behind to do something else. So the big element of control that they all retain is the freedom to choose whether or not to do it. To take that point to absurdity, Kelvin would find no thrill in extreme proximity to death in old age. Hardened arteries somehow don't count as edgework. So choice is a big factor in attitudes to danger. The fright factor[14] of involuntary risk, or risk that people can't escape – exposure to pollution is another good example of a risk we can't avoid – often feels far worse than risks we choose, such as adventurous sports. Surfers Against Sewage is a real organisation, with a fair point about which risks are tolerable. Base-Jumpers Against Climate Change – if it existed – wouldn't be altogether daft either.

If parachuting were as safe as a day at the office, would it have the same thrill? For many, yes. The allure – for those who feel it – is not solely the risk as it appears in the statistics, but the risk as it feels in the moment. Instinctive fear of being high up is probably an evolutionary plus. But evolution hasn't caught up with parachutes yet. In other words, danger isn't measured only in the head, it's measured in the skin. As with flying or roller-coasters, we can feel afraid, even enjoy the fear, with only a little more than usual to worry about.

Objective measures of harm are not irrelevant to this subjective thrill. If you knew the fairground ride was a death trap, you wouldn't get on. So we say to ourselves, 'I know I'll be all right, but I don't feel it', and with that careful balance enjoy the ride. For many others an old saying has to be taken seriously: risk is its own reward. Just don't be surprised when people choose different levels of risk for different aspects of their lives. Attitudes to danger vary even within individuals, for any of whom risk is strangely, but not necessarily foolishly, compartmentalised.

LIFESTYLE

NORM LAY IN BED, eyes shut, listening to his body. He was middle-aged and felt it. His body told him that it hurt.

He had recently taken up exercise after studying the data on chronic illness over lunch with Kelvin: a burger for Kelv, a salad for Norm. Running was his attempt to slow the raindrop as it trickled down the window. Although there were other consolations too. For example, beer tasted better afterwards, bitter with the sweet taste of righteousness.

So for twenty minutes each day on 200mg of flecainide acetate to control his cardiac arrhythmia, paunch creeping over the band of his shorts, he forced his feet down lanes lined with brambles or through damp leaves by the canal, blowing hard at the lid of his closing coffin, then sucked the air back like a glutton. He heaved his carcass all the way up to jog speed, nearly, until the aches were washed out by adrenaline. This was what it was all about. This was striving.

'Commit, Norm, commit!'

He pushed again until his body screamed that it was alive and feeling too much, all to counter the long hours when it seemed to feel nothing but fed up. He drove his thin legs and squeezed wheezing lungs to prove his frame could raise one more militant shout against age and wouldn't be dried and silenced. When he was running, he was living; it was transcendent and for ever, for 22 minutes. And Norm knew precisely how much extra life 22 minutes of exercise was worth, on average.

He finished. He slumped, while every muscle said it was spent and he wondered if he could have sustained his Mo-Farah finishing burst from the gate into the allotment instead of waiting for the lane. And he looked at his watch and he saw – and sighed – that he was, after all, at 22 minutes and 18 seconds, slower. Now he could go to the pub.

LIFESTYLE IS A new kind of danger to Norm. The hazards he's lived with so far have been instant, like violence or accidents, the kind that hit us over the head with a swift goodnight. But Norm is now running from a more sinister threat, another type of mortal hazard with slower effects that go stealthily into the blood one cancerous bacon butty or poisonous pint at a time, potential killers by degrees that might catch up with us later in life, as something surely will.

The first mortal hazard – the quick one – is known as acute risk; the second is chronic. Murder with a chainsaw is an acute risk, obesity a chronic one that takes time to do its worst. Of course, the same hazard might be both: too much booze can do you in quickly when you fall under a bus, or slowly stew your liver. But in general it helps to separate them.

So far, we've used the MicroMort to describe acute risks. For chronic risks like obesity or Norm's current long-term lifestyle anxieties, we introduce a little device we have called the Micro*Life*. Here's how it works.

Imagine the duration of your adult life divided into 1 million equal parts. A MicroLife is one of these parts and lasts 30 minutes. It is based on the idea that as young adults we typically have about 1 million half-hours left to live, on average.*

Sounds unimpressive. But we like the MicroLife. It is a revelatory little thing. Like the MicroMort, it brings life down to a micro level that's easy to think about and compare: half-hour chunks, of which we have 48 a day. Think of it as your stock of life to use up any way you choose,

* Life-expectancy for a man aged 22 in the UK is currently about 79, or a further 57 years, which is 3,418,560 minutes, or 20,800 days, or 500,000 hours, or 1 million half-hours. For a woman, life-expectancy is about 83, so her million half-hours start at 26. This will not be true for everyone, but short of clairvoyance it'll do, roughly, more or less, overall.

1 million micro bits of a whole adult life, each worth half an hour, yours to spend. Watching the Eurovision song contest? Bang, 6 MicroLives gone, just like that, never to have again.

So the simple passing of time uses up MicroLives. Every day we get up, move around, stuff tasty things into our bodies, discharge smelly things out of our bodies and go to bed – perhaps with the thought, if we're gloomy, that there go another 48 MicroLives from our allotted span.

But extra MicroLives can also be used up by taking chronic risks. So although time passes to its own beat, our bodies can age faster or slower according to how we treat them. If we jump around more, and stuff less or better, how much can we slow the steady tick-tock towards disease, decrepitude and death? And if we indulge and slob, how fast might our own clocks run?

In other words, MicroLives can measure how fast you are using up your stock of life, faster or slower depending on the chronic risks to which you're exposed. If your lifestyle is chronically unhealthy, you'll probably burn up your allotted MicroLives that much quicker, and die sooner, on average.

For example, lung cancer or heart disease often follows a lifetime's smoking, and subsequently reduces life-expectancy – again not for everyone, but overall. Some people seem indestructible, smoke like a chimney and drink like a fish and never look the worse for it. But, on average, even if chronic risks don't kill you straight away, they tend to kill you sooner than if you had avoided them. Again, if we count the bodies, we can estimate how many years are lost overall, whether to obesity, smoking or sausages, and we can convert this loss of life into the number of MicroLives burned up by unhealthy living. Thus, exposure to a chronic risk equal to 1 MicroLife shortens life on average by just one of the million half-hours that people have left as they enter adulthood.

It turns out that one cigarette reduces life-expectancy by around 15 minutes on average, and so two cigarettes costs half an hour, or 1 Micro-Life. Four cigarettes are equal to 2 MicroLives.

About two pints of strongish beer is also 1 MicroLife. Each extra inch on your waistline costs you around 1 MicroLife every day, seven a week, about 30 a month and so on. According to recent research, so does watching two hours of TV. An extra burger a day is also about 1 MicroLife. We'll reveal the calculations behind these risks in a moment.

We could simply add up all these MicroLives, half-hour by half-hour, to see roughly how much time on average you lose in total from whatever your lifespan might have been. But the end of life is often far away, like the end of the story, and a lost half-hour deferred until you are in your dotage hardly seems to count. As a media doctor said: 'I would rather have the occasional bacon sarnie than be 110 and dribbling into my all-bran.' But by thinking of exposure to chronic risks like an acceleration of the speed at which you use up your daily allotment of MicroLives, we can do something more vivid and immediate. We can show how much your body ages each day according to the chronic or lifestyle risks that you take.

Ordinarily, remember, we use up 48 MicroLives (MLs) a day. But remember, too, that smoking four cigarettes burns an extra 2 ML. So if you smoke four cigarettes in a day, you've used not 48, but 50 ML that day. In other words, after a 24-hour, four-fag day, we could say that you are 25 hours older.

And that's not a bad representation of what can happen biologically Bodies do often age faster when we do bad stuff to them. Twenty cigarettes daily means you burn an extra 10 ML a day on average, or become 29 hours older with every 24 that pass, or move towards death five hours faster, every day.

Suddenly, chronic risk feels a lot more here and now than the far-away payback that we typically put off thinking about until we're spent anyway. It counters what is sometimes known as temporal discounting. By bringing chronic risk down to the small scale of what happens today, rather than thinking of it only as a life-size problem deferred, the Micro-Life makes chronic risk a good deal more real and immediate.

But should it? You might argue with this. You might prefer to put off facing up to your lifestyle risks. You might argue that you shouldn't be confronted with the payback until it actually occurs, late in life. On the other hand, it could be argued that the damage is done now, at the point of consumption, so we should measure it now.

MicroLives allow us to make simple comparisons of chronic risk, just as we did for acute risk with MicroMorts. Now we can compare sausages with drinking or smoking, X-rays with mobile phones. We can compare a CT scan with watching the Hiroshima atomic bomb from the Hiroshima suburbs, getting fat with getting fit, unprotected sex with unprotected sun – and we will. In Chapter 27, Figure 37 shows a selection of MicroLife hazards.

If calculations terrify you, move swiftly on to the next chapter. Because next, we'll find out exactly how these calculations are done and what the evidence is for the MicroLife costs and benefits we've identified. Treat it as a statistical detective story.

We begin with the 1 ML cost of an extra burger every day, mentioned earlier. This was reported in the *Daily Express* in a story about the dangers of red meat, based on a study from Harvard University.[1] The *Express* said: 'If people cut down the amount of red meat they ate – say from steaks and beefburgers – to less than half a serving a day, 10 per cent of all deaths could be avoided.'

Oh to be one of the 10 per cent for whom death could be avoided! But this is not what the study said. Its main conclusion was that an extra portion of red meat a day – this being a lump of meat around the size of a pack of cards or slightly smaller than a standard quarter-pound burger* – is associated with a 'hazard ratio' of 1.13: that is, a 13 per cent increased risk of death. Put aside any doubts about the validity of this number for a moment and take it at face value. What does it mean? When our risk of death is already 100 per cent, surely a risk of 113 per cent is an exaggeration?

Let's consider two friends – whom we'll call Kelvin and Norm – who are both aged 40, and just for the moment let's make the unrealistic assumption that they are pretty much alike in most lifestyle respects that matter, apart from the amount of meat they eat.†

---

* Actually 85 grams, or about 3 oz.

† The assumption is – in line with the Harvard study – that they have the same average weight, alcohol intake, exercise regime and family history of disease, but not necessarily quite the same income, education and standard of living. This is how the Harvard team analysed the risks, by trying to focus on the effect of the meat we eat without too many other factors in the way.

Carnivorous Kelv eats a quarter-pound burger for lunch from Monday to Friday, while Normal Norm does not eat meat for weekday lunches, but otherwise has a similar diet to Kelvin's. We are not concerned here with their friend Particular Pru, who has given up eat meat and turned veggie after reading the *Daily Express*, but who might succumb to a contaminated sprouted fenugreek seed.

Each person faces an annual risk of death, the technical name for which is their 'hazard' or, somewhat archaically but poetically, their 'force of mortality' (for a fuller discussion of the force of mortality, see Chapter 26, 'The End'). A 'hazard ratio' of 1.13 means that, for two people like Kelvin and Norm, similar apart from the extra meat, the one with the risk factor – Kelvin – has a 13 per cent increased *annual* risk of death – not an overall risk, obviously – during a follow-up period of around 20 years.

This does not imply that his life will be 13 per cent shorter. To work out what it really means we have to go to the life-tables provided by the Office of National Statistics. These tell us the risk that an average man – Norm, say – will die at each year of age. In 2010 this risk or hazard was at its lowest for those aged seven (see Chapter 2, on infancy), at 1 in 10,000: it then rises to 1 in 1,000 at age 34, then to 1 in 100 at age 62, until at age 85, 1 in 10 will die before their 86th birthday. So very roughly the annual chance of death increases tenfold around every 27 years, which works out at doubling every nine years, or about a 9 per cent extra risk of dying before the next 12 months are out, for every year that we are older. The tables also tell us life-expectancy at any given age, assuming the current hazards, and, having survived to age 40, Norm is expected to live a further 40 years, until he is 80.

From this we can work out Carnivorous Kelv's prospects by multiplying all Norm's hazards by 1.13. After a little work in a spreadsheet, we find that Kelvin can expect to live 39 more years on average, a year less than Norm. So Kelvin's lunch – if he eats the same lunch all his life and if we believe this hazard ratio – is associated with the loss of one year in expected age at death, from 80 to 79.

Is that a lot? Kingsley Amis said, 'No pleasure is worth giving up for the sake of two more years in a geriatric home at Weston-super-Mare.'[2] It is for readers to decide. But we cannot say that precisely this amount of time will be lost. We cannot even be very confident that Kelvin will die first. In fact, there is only a 53 per cent chance* that Kelvin will die before Norm, rather than 50:50 if they eat the same lunch. Not a big effect.

But it sounds rather more important if we say that this lost year (1/40th of the remaining life) translates very roughly to one week a year, or roughly half an hour a day. So one extra MicroLife burned up for each daily burger. So, unless you're a very slow eater, you expect to lose more life than the time it takes to eat your burger.

But we can't even say the meat is directly causing the loss in life-expectancy, in the sense that, if Kelvin changed his lunch habits and stopped shovelling down the burgers, his life-expectancy would definitely increase. Maybe there's some other factor that both encourages Kelvin to eat more meat and leads to a shorter life.

Income could be such a factor – poorer people in the US tend to eat more burgers and also live shorter lives, even allowing for measurable risk factors. But the Harvard study does not adjust for income, arguing that the people in the study – health professionals and nurses – are broadly doing the same job. We think that many of these studies about diet should be taken with a pinch of salt (although perhaps not too much, since it may increase your risk of heart disease).

We can also look deeper into the calculation for other bad behaviours that receive a finger-wagging. So, next, smoking. The evidence against smoking is much better than that against red meat. In terms of shortening your life, a very basic analysis estimated a 6.5 years difference in life-expectancy between smokers and non-smokers, which is 3,418,560 minutes.[3] They considered median consumption of 16 cigarettes a day from ages 17 to 71, which comes to 311,688 cigarettes. Making the simplifying assumption that each cigarette contributes equally to the risk, this comes to 11 minutes loss in life-expectancy per cigarette, or around 3 cigarettes for a MicroLife.

---

* If we assume a hazard ratio h is kept up throughout their lives, then some rather elegant maths tells us the probability that Kelvin dies before Norm is precisely $h/(1+h)$, which when $h = 1.13$, = a 53 per cent chance Kelvin dies first, rather than 50:50 if they eat the same red meat.

Figure 20: **Some MicroLives: how many gained or lost for different activities**

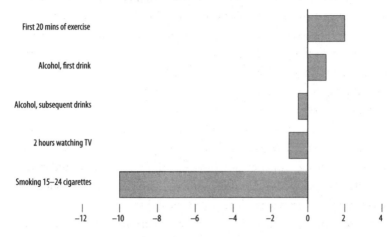

This simple analysis only compares smokers with non-smokers, who may be different in all sorts of ways that could also influence their health. A more sophisticated analysis would consider the effect of stopping smoking, and fortunately that's been done in a classic study on 40,000 UK doctors, many of whom gave up smoking during the period of the study from 1951 to 2001.[4] They estimate a 40-year-old man who stopped smoking gained 9 years in life-expectancy, or 78,000 hours in total, and from this we can estimate around 2 cigarettes for a MicroLife. So 2 ciggies is roughly equivalent to 1 burger, when taken daily.*

So what about the booze? The precise effect of alcohol on all-cause mortality is controversial since, although it can cause accidents (particularly for drivers and young binge-drinkers), give you liver disease and increase the risk of some cancers, it can also protect your heart. So the 'dose-response' curve for annual risk is J-shaped in middle age, meaning

---

* If he carries on smoking, he is only expected to live another 30 years or 11,000 days, so he is on average losing 7.2 hours per day (14 MicroLives): we can imagine him speeding recklessly towards his death at 31 hours a day. During these 30 years he might smoke 325,000 cigarettes (assuming the higher consumption of 30 a day in the 1950s and 1960s). This works out at 15 minutes lost per cigarette smoked.

that the risk falls slightly with the curve of the 'J' as you consume a small amount of alcohol, then rises as you consume more. Very roughly, it looks like the first drink each day adds a MicroLife, but extra drinks take it away again, and more. So the first is medicine, the second poison, the third poison and so on. It does *not* go medicine/poison/medicine/poison ... – that would be ridiculous, and would mean that one would need to have an odd number of drinks.[5]

All this is depressing, but what about the benefits of a good diet and hearty exercise – eating muesli and running (although not at the same time)? As A. A. Milne said, 'A bear, however hard he tries, grows tubby without exercise', and we can try and work out what that tubbiness might do to your life. A recent review estimated a risk * which translates to around 1 ML lost for every day that you are 5 kg overweight.[6] Seriously obese people can lose ten years off their life-expectancy, similar to smokers.

The European Prospective Investigation into Cancer and Nutrition (EPIC) study in Norfolk compared people who ate their five portions of fruit and veg a day with those who didn't (and the researchers checked their honesty about their diet by measuring vitamin C levels in their blood).[7] The hazard ratio was 0.69, showing around 30 per cent less annual risk associated with a better diet, or around 3 ML a day saved, as if they were only ageing 22.5 hours instead of 24.

And all that running around? The UK guidelines recommend we all take 30 minutes moderate or vigorous activity on five days each week – a total of 2.5 hours per week or 22 minutes a day. When asked in 2008 about their physical activity, 39 per cent of men and 29 per cent of women claimed they did this much exercise each week.[8] However exercise, like sex, tends to be over-reported, just as alcohol is under-reported – if we believed what people say they drank, half the alcohol bought in the UK must be left in the bottle or chucked down the sink.

---

* The study gives an estimated hazard ratio of 1.29 for all-cause mortality per 5 kg/ m^2 increase in body mass index (BMI) over the optimum of 22.5 to 25 kg/m^2. For a man/woman of average height (1.75m/1.62m), this corresponds to a hazard ratio of around 1.09 per 5 kg overweight, translating to one MicroLife per day.

Which we doubt. When people carry accelerometers that measure their true exercise, only 6 per cent of men and 4 per cent of women meet the government-recommended levels of activity. We are part of a population that is both tubby *and* deluded.

A huge review of 22 studies involving nearly 1 million people concluded that 2.5 hours a week of 'non-vigorous' activity was associated with a hazard ratio of 0.81 compared with being a complete slob – a 19 per cent reduction in annual risk of death.[9] This works out as about an hour per day, or 2 ML, added to life-expectancy for an average of 22-minutes-a-day activity, which must be why that's the length of Norm's run – quite a good return for the investment of getting off the sofa.

So it's a good idea not to be a couch potato, and a Swedish study even showed that it's never too late to start, with increasing physical activity in middle age eventually bringing your risk down to the level of people who have always been active – similar to the benefits of giving up smoking.[10]

Some naive extrapolation would suggest that if we do lots of exercise we could live for ever, but sadly there is a strong law of diminishing returns. People who do 7 hours moderate exercise a week, a full hour per day, only reduced their risk by 24 per cent, equivalent to about 1.5 hours added to their life for each day of exercise.

So, very roughly, compared with doing no physical activity, the first 20 minutes or so per day pays rich dividends, but any more than that is approximately pro rata: 20 minutes of exercise earns you 20 minutes of added life-expectancy. It's like time stops for you when you're exercising. And on a treadmill it can certainly feel that way.

Our circumstances when we're born – which we can't do much about – can also be expressed in MicroLives. For example, being female rather than male (worth an extra 4 ML, or two hours per day), being Swedish rather than Russian (21 a day for males, or more than ten extra hours) and living in 2010 rather than 1910 (15 a day, or 7.5 extra hours).

Of course, working out the health effects of lifestyle is not a precise art. It is impossible to put exact numbers on the harm we do to our bodies with an extra cigarette, sausage, pint of beer or glass of wine, or the lack of five-a-day fruit and veg, or – on the virtuous side – the benefits we might reap on the exercise bike or by being otherwise wholesome and pure.

But we can calculate the effects approximately, by averaging over many lifetimes, and it is worth trying, especially in a world of exhortations to self-improvement or stories of how you too can stay forever beautiful and young – or not.

There is one big difference between MicroMorts and MicroLives. If you survive your motor-bike ride, your MicroMort slate is wiped clean and you start the next day with an empty account. But if you smoke all day and live on pork pies, then your used-up MicroLives accumulate. It's like a lottery where the tickets you buy each day remain valid for ever – and so your chances of winning increase. Except that in this case you really don't want to.

People spend a fortune on cigarettes, but what would you sell half an hour of your life-expectancy for? Governments put a value on MicroLives, as they do with MicroMorts. The UK National Institute for Health and Clinical Excellence (NICE) has guidelines that suggest the National Health Service will pay up to £30,000 if the treatment is expected to prolong life by one healthy year. That's around 17,500 ML. This means that NICE prices a MicroLife, or half an hour of your remaining life-expectancy, at around £1.70, almost exactly what the Department of Transport says it would pay to avoid a MicroMort.

Does this mean the government ought to pay you £1.70 for every two cigarettes you resist or for every day that you keep that extra inch off your waistline? Possibly, except that it wouldn't work like that: we could all claim to resist hundreds of cigarettes a day, non-smokers especially.

Now that Norm is into middle age and full of regret for the burgers that slipped through in earlier life,* puffing hard to put time back in the bank, it's interesting to note that he hasn't become pure, merely smug. A 22-minute run might buy him an extra hour of life, but one effect is that he feels happier about spending it in the pub.

His behaviour is known as risk compensation, related to the idea in

---

* Kelvin? Il ne regrette rien.

Figure 21: **Scandinavian orange juice**
But if you live longer than your friends, who will you talk to?

Chapter 15, on transport, that we all have an in-built risk thermostat. Taken your vitamins? Great. Eat extra chips!

There's experimental evidence to support this.[11] In one trial, in a culture of heavy smokers, some participants were given pills and told they could have a cigarette break. All the pills were placebos. But those who thought they'd taken a vitamin pill were far more likely to go for a smoke (89 per cent to 62 per cent). In another study surveying people's sense of vulnerability, those who took vitamin pills somehow thought they were less likely to be hurt in an accident.

So if you've banked some health, you might feel free to spend it. How these attitudes and behaviours shake out – some beneficial, some compensatory and harmful – takes some calculating, except for vitamin pills, which don't do most people much good anyway. But it does suggest that the risk calculation for healthy behaviour also needs to take account of the cream cake afterwards.

A summary of the research on exercise suggests that it still brings substantial benefits, although less weight loss than you might hope, perhaps because people eat more.[12] The research does not suggest that exercise leads people to eat so much more that they become fatter on average, contrary to much online gossip. But do mind that you don't seek too much cake compensation for all your hard work.

18

HEALTH AND SAFETY

NEARLY HOME, Norm saw a line of plastic bollards outside his house where the road swerved up and over a small hill. At either end was a traffic light, in the middle a van and generator. On the back of the van was written 'Urgent Response Vehicle'. Beneath this: 'Limited to 56 m.p.h.'. Next to the van stood a man studying a plan.

'What's up?' said Norm.

'Two-way traffic control, sir.'

'Oh right. Because ...?'

'Bollards. Can't have a lane closure without traffic control. Asking for a head-on.'

'Right, so the traffic lights are for the cones. And the cones ...?'

'Protect the traffic lights.'

'... ?'

'Well we can't have lights just stood there in the road like that without diverting the traffic, can we?'

'So the cones are for the lights and the lights are for ... This might sound stupid, but can't you just ...?'

'Remove it all? No sir. Traffic lights are traffic control. Bollards is protection. Without one the other's a hazard, obviously.'

'You're not digging up the road, then?'

'Why would we do that?'

'Right. And you are, if you don't mind ...?'

'Fire regs compliance. At present, the site lacks an assembly point.'

'Assembly point ... for ...?'

'Self and Eric.'

He pointed to another man reading a newspaper.

'A designated offsite site-assembly point is required to be signified –
in the field there, for instance – where in the event of fire or explosion or
other cause leading to site evacuation we can establish that all personnel
are accounted for. Mind that cable, sir.'

'You're ensuring that your own presence is compliant?'

'Exactly.'

'To count two of you?'

'Indeed.'

'When one of you does the counting?'

'Meaning?'

'Well, why not just *not* come?'

'You some kind of Marxist, sir?'

'No, I mean just don't come. Then you won't run the risk of being
non-compliant.'

'Then how would we know?'

'Know what?'

'If it was compliant for inspection?'

'Right ... er... gotcha ... Thanks.'

'Just serving the community, sir.'

DOES THE STORY of Norm and the bollards ring true – everything you
always thought about health and safety? Funny, isn't it, how we love a
story if it confirms that we're right?

We're good at filtering evidence, as we saw in Chapter 10, on Big
Risks. So much so that we'll sometimes believe stories that are absurd –
provided they fit our preconceptions.

Like the story that throwing sweets into the audience at pantomimes
is banned – to take one daft example. Except that it's not.[1] Rumours otherwise are just a popular myth that Britain's Health and Safety Executive (HSE) tries hard to dispel, but people are too willing to believe.
Conker fights are OK too: 'Realistically the risk from playing conkers is

incredibly low and just not worth bothering about', says the HSE. But the myths persists. Loony laws are too good a story.

So for all its efforts, the HSE's reputation is probably sealed by the stories about it. 'Elf and safety' is a curse, a joke. In the league of comic villains it beats the mother-in-law.

'One of the most well-worn and dispiriting phrases in the English language,' said Judith Hackitt, chair of ... the Health and Safety Executive.[2]

And now and then, in the name of health and safety, strange things do get said, which hardly help the myth-busting. Take John Adams, who was told the windows in his block needed coating with plastic lest people walking below during a storm, as they do, obviously, at just the spot and moment a window blew out, copped a shower of shards.[3]

Official statistics record two deaths that year from glass injury. They don't say whether these were caused by windows shattering in a storm. It seems unlikely. When Adams asked if there had been a risk assessment, he was told 'it could happen', which we suppose is an assessment of sorts. Meanwhile, the lift in his seven-floor block was out of order for long stretches. There were 634 deaths on stairs that year. Norm's story is in a long, absurdist tradition with just enough evidence to make it almost plausible. In popular demonology, Health-and-Safety-Man might as well be a villain on *Doctor Who*.

Psychologists talk of a cognitive bias for zero-risk, a preference for the certain elimination of a small risk over the probable reduction of a greater one. After all, what is safety but the absence of danger? So if you believe in a safer world, picking on risks you think you can cut out altogether may seem to make a kind of sense, almost whatever it takes.

Hearing all this, health and safety sounds like the preserve of zealots. Mind you, it can be equally tempting to see the other side the same way, as *Daily Mail*-style 'elf-an-safety-bureaucracy-gone-mad' ranters who fancy themselves freedom fighters. Any chance of moderation?

Because, needless to say, there's plenty of room in the middle of this argument. After all, what would you do – fit plastic to the windows, mend the lift, both or neither? We'd mend the lift, though not only for safety. We guess most people would do the same. Funnily enough, the Health and Safety Executive itself is often on that same middle ground, while it is others who whip up wild health and safety fears to stop things they don't like (see Chapter 9, on drugs, for the anthropologist Mary Douglas's ideas of risk as a form of social control).

Not all health and safety is a joke. Nor are those who campaign about it necessarily fighting over trivial, residual risks. *Hazards* magazine – another clue in the name – argued in 2012 that the HSE was ignoring evidence of an occupational cancer epidemic (see pp. 200–201) that kills 15,000 people a year. The magazine is free with phrases such as 'corporate killers'.[4]

So although irked by its extremes and absurdities, Norm has reason to be grateful to health and safety, especially for a long-term shift in attitudes to workplace accidents. Even in 1974, when the HSE was formed, 651 employees were recorded as killed at work, an average risk of 29 MicroMorts per year. By 2010 this had fallen to 120, equal to 5 Micro-Morts per year, an 82 per cent decrease.[5]

Self-employment is more dangerous. Fifty-one self-employed people were killed in 2010, equivalent to 12 MicroMorts per year, more than twice the risk for employees. Injuries have also fallen dramatically: days off work through injury and illness are down by about a third in the past ten years alone.[6]

Britain also fares well in comparison with other EU countries: excluding road transport deaths, workers in Britain were on average exposed to 10 MicroMorts per year, compared with 17 in France, 19 in Germany, 26 in Spain, 35 in Poland and 84 in Romania,[7] according to Eurostat. Tempting though it is to say that this is because the riskiest job anyone has in modern Britain is standing in a shop or sitting at a computer, Britain manufactures about as much as France (though much less than Germany).

In the US, the Bureau for Labor Statistics provides an extraordinary glimpse of the fate of 130 million workers in 2010.[8] It records that 4,547 were killed, a rate of 35 MicroMorts per worker per year. The most common cause was highway accidents, which are excluded from the European figures. Without them, the US rate is around 28 MicroMorts per year, about the same as Spain.

The second most common cause of death in the US, larger even than falls, is 'assault and violent acts', comprising 18 per cent of all workplace fatalities. This included 506 homicides (down from 860 in 1997). So each year US workers face on average around 4 MicroMorts' risk of being murdered at work, not much less than the average risk to UK workers from all causes. Perhaps they need body armour rather than safety helmets.

Reliable statistics about the wider world are harder to find. For example, India reported 222 fatal accidents at work in 2005, which was rather an understatement according to the International Labour Organisation (ILO), which reckoned the true number nearer 40,000.[9]

Among 2 billion workers worldwide in 2008, the ILO estimated 317 million injuries requiring more than four days' absence, and 320,000 people killed at work.[10] Only 22,000 of these deaths were reported through official channels. The rest are a 'guesstimate', giving an average of 160 MicroMorts per year per worker, compared with 6 in the UK.

All these figures are averages. They include armies of workers toiling over computers, which might cause stress, repetitive strain and back injury but are seldom fatal, and which tend to pull the average mortality risk down. Other occupations pull it up. The history of coalmining, for example, is spotted with disaster, such as the 1,099 miners killed at Courrières in France in 1906, or the gas explosion that killed 439 in Senghenydd, Wales, in 1913. Behind these terrible events there was a steady stream of fatalities in smaller accidents, with that frightening regularity of the apparently unique and unpredictable.

Records of mining accidents began in the UK in 1850, and since then more than 100,000 coalminers have been killed and hundreds of thousands injured or suffered illness. The years 1910 and 1911 were two of the most dramatic, with violent confrontation between miners and mine-owners. Following strikes and rioting in South Wales, Winston Churchill sent in the troops. There were 1,308 deaths among 1,100,000 miners in 1911[11] – that is 1,190 MicroMorts per year, or around 5 per shift, as if every miner on every shift went sky-diving every other day.

The Coal Mines Act of 1911 set up rescue stations and was intended to improve safety. Even so, around 1,000 miners were killed year after year, and in 1938 the tally was still 858 deaths, or 1,100 MicroMorts a year. Following nationalisation in 1947, mine safety steadily improved, until by 1961 there were 'only' 235 deaths in accidents, a risk, after taking account of the declining workforce, of 400 MicroMorts a year, or around 2 per shift. This improved further, but even though there are now only 6,000 coalminers in the country,[12] a spate of accidents in recent years has brought a return to the fatality rates of the 1960s: in 2005–10 the fatality rate was reported to be the equivalent of 430 MicroMorts per year. Privatised mines were accused of cutting corners.[13]

Coalmining accidents elsewhere are repeatedly in the news. Forty-eight miners were killed in Pakistan in 2011. In China economic growth is driven by coal, but it is deeply buried at an average depth of 400 metres. Even according to official figures there is a high death toll: around 250,000 since 1949, around 7,000 in 2002 and 2,600 in 2009.[14] Assuming 4 million miners underground, that's around 650 Micro-Morts a year (official figures), which was about the risk in Britain in 1950.

Unofficial estimates put the current total nearer 20,000, or 5,000 MicroMorts a year.[15] This may be unsurpassed as a modern occupational risk. It is worse than Britain 150 years ago. Vast numbers of small mines are run by corrupt local officials under minimal central control. The Asian Development Bank [16] has commissioned worthy reports, but attempts to impose safety regulations or closures often lead only to the growth of illegal mines.[17]

The highest-risk occupation in the UK today is commercial fishing. A recent study recorded 160 deaths in the UK between 1996 to 2005, which works out as 1,020 MicroMorts per year per fisherman.[18] Fourteen died in coastal shipwrecks, 59 drowned when their boat sank or capsized, mainly because it was unstable, overloaded or unseaworthy. About half the fatalities are solitary fishermen, not helped by a culture of shunning personal flotation devices. Perhaps unique among industries, these risks have not declined since the Second World War, although before that they were even higher: 4,600 MicroMorts per year in 1935–8, close to the unofficial risk for Chinese miners today.

Similar rates of 1,000–1,160 MicroMorts per year apply to Alaskan, Danish, French and Swedish fishing fleets, while the risks are even higher in New Zealand (2,600). Besides commercial fishing, other high-risk UK occupations between 1996 and 2005 include dockers (280 MicroMorts per year), refuse and salvage workers (250) and agricultural machinery drivers (180).

Of course, it is not just workers who can suffer from industrial accidents – bystanders are caught up too. A famous example was the great London beer flood of 17 October 1814, when huge vats of porter burst in the Meux brewery on the corner of Oxford Street and Tottenham Court Road.[19] More than a million litres of strong beer – enough to fill 20,000 barrels – broke through brick walls, demolished two houses, seriously damaged the Tavistock Arms pub and poured into cellars occupied by the local poor. Nine people drowned, and the fourteen-year-old barmaid Eleanor Cooper was crushed by rubble in the Tavistock Arms. Others, of course, rushed to the scene to fill up pans and kettles. The story that one fell victim to acute alcohol poisoning could be an urban myth. In the later court case the disaster was ruled to be an act of God, and the brewery escaped responsibility.

Just over a century later this bizarre event was eclipsed by the Boston molasses disaster of 15 January 1919.[20] Like a bad disaster movie, a huge tank of 8 million litres of molasses burst – equal to around three Olympic swimming pools – and a black, sweet and sticky tsunami, 3 metres high and travelling at 35 miles per hour, swept out, demolished buildings, damaged the overhead railway, drowned numerous horses, killed 21 people and injured 150 more. The United States Industrial Alcohol Company was found responsible and paid compensation.

If there's black comedy in these stories of absurd accidents, events in Bhopal in 1984 are of a different order. Thirty tonnes of methyl isocyanate (MIC) leaked from a tank belonging to a subsidiary of the US Union Carbide Corporation.[21] MIC is a gas, but heavier than air. There was no escape from nearby shanty towns. Three thousand died immediately. Recent estimates suggest a final death toll of 25,000 and half a million injuries, many permanently disabling. The court cases continue.

The victims of Bhopal illustrate the distressing long-term effects of industrial accidents and exposures. The ILO estimates that 2 million people died from diseases caused by their jobs in 2008. In the UK the

HSE estimates that in 2009 there were around 8,000 cancer deaths due to people's previous occupations, half of them from asbestos.[22] While deaths from pneumoconiosis caused mainly by mining are now falling – from 453 in 1974 to 149 in 2009 – deaths from asbestosis and mesothelioma (cancer caused by asbestos) are rising, and not expected to peak until 2016.

This is a chronic risk rather than an acute one. When exposure to asbestos kills, it usually does so long after the event. It would be useful to know the chronic risk faced by asbestos workers or miners in terms of lost MicroLives per day of employment. But something strange happens in this calculation. The long-term mortality of asbestos workers turns out to be only 15 per cent higher than the general population,[23] roughly corresponding to the loss of around 2 MicroLives per day, no worse than a few cigarettes. Even more paradoxical is the finding that the mortality of 25,000 British coalminers was 13 per cent *lower* on average than men living in the same regions.[24]

The explanation for these findings is known as the 'healthy worker effect'. Men would not be down the pit unless they were fit and healthy, and so, compared with average people, miners can appear to have better survival in spite of their exposure to coal-dust, because they are drawn from a healthier population. This makes it hard for epidemiologists to calculate the harm of poor working conditions except in general terms – or by counting bodies.

We've seen how some occupations are still risky, although not at Victorian levels. But how high must the risk be before the HSE decides that something should be done? The HSE's philosophy is based on what is known as the 'Tolerability of Risk' framework, and can be nicely converted to MicroMorts.[25]

Potential hazards are thought of on a spectrum of increasing risk, divided into three loosely defined categories. At the top are 'unacceptable' risks: whatever the benefits of the activity, something must be done to protect either the workers or the public, or both. At the bottom are 'broadly acceptable' risks: not quite zero but considered insignificant, the sort of thing we would regard as normal in our daily lives.

In between these extremes lie the 'tolerable' risks: those we might be prepared to put up with if there is sufficient benefit, such as providing valuable employment, personal convenience or keeping the infrastructure of society going: somebody has to do the dirty work. The HSE makes clear that risks should be considered tolerable only if they are carefully weighed up using the best evidence, are periodically reviewed and are kept 'as low as reasonably practicable', a set of criteria known as ALARP. (The HSE has other great mnemonics that are not so easy to pronounce: duties to reduce risk may be SFAIRP – 'so far as is reasonably practicable'.) But how does anyone decide what is unacceptable, broadly acceptable or tolerable? The HSE suggests some rough rules of thumb.

First, it states that an occupational risk might be considered 'unacceptable' if the chance of a worker being killed is greater than 1 in 1,000 per year, or 1,000 MicroMorts per year. On this metric, the risks faced by pre-nationalisation British miners would now be considered 'unacceptable', and current commercial fishing is around this level. The HSE exclude 'exceptional groups': presumably serving in a war zone counts as exceptional, for example when the average risk faced by 9,000 servicemen and women in Afghanistan reached 47 MicroMorts a day, or around 17,000 a year in late 2009.[26]

For members of the public, rather than employed workers, the HSE considers a 1-in-10,000-a year-risk – that's 100 MicroMorts a year – as generally unacceptable. At the other extreme, risks are considered broadly acceptable if they are less than 1 in a million per year – 1 Micro-Mort – the current estimate of the average risk of being killed by an asteroid.

Even such a minimal risk, if applied to the whole population of the UK, would mean around 50 deaths a year. This brings another curiosity. Imagine the headlines if 50 deaths happened at once. As we've seen when discussing railway disasters in Chapter 15, cold MicroMort calculations are easily trumped by 'societal concerns' about disasters with multiple fatalities, hazards that affect vulnerable groups such as young children, or risks imposed on people just because of where they live.

The HSE offers separate guidance that reflects this: the risk of an accident involving 50 deaths should be less than 1 in 5,000 per year.

Spread across a population of more than 10,000, it is less than 1 Micro-Mort a year each. From an individual perspective this might be considered 'acceptable', but because we don't like disasters (although people clearly love reading about them in newspapers), huge amounts of money are spent to make tiny risks even tinier.

19

RADIATION

Prudence scanned the shelves of vitamins, tonics and remedies in the chemist's shop. Pansy swung from her mother's arm. Above them in a bundle tied by their feet, soft toys wore the label 'I'm a sloth'.

'What are those mummy?'

Prudence looked up.

'Sloths.'

'Can I have a sloth?'

'No.'

'What do they do, mummy?'

'Nothing.'

'Why not?'

'They just sit there.'

'Please mummy?'

'Stuffed.'

'Pleeease?'

'Doing nothing.'

'Can I?'

'Like your father.'

Prudence disapproved of soft toys. Think of the dust mites and plastic eyes. She disapproved of her husband's inertia too. His cough, for instance.

She took from the shelf a small bottle of 37 life-giving essential vitamins, minerals and nutrients for health and vitality, including active antioxidant extracts of Korean Ginseng, and tossed it into the basket with two others of high-strength organic sunscreen.

'It's a cough,' he said that evening, as they sat watching the news about an accident at a nuclear power station while Prudence ate her ten o'clock banana.

'Can't a man have a cough and it just be a cough?' he said, then added, mysteriously: 'Fires sometimes go out.'

But the story on the news was of a fire that didn't go out. 'Meltdown,' they said and talked of criticality excursions, which sounded like a fatal bus ride. Why he agreed with it, in his boys-and-lasers way, she'd never know. The boffins were clueless, it wasn't natural, and she'd rather live in a mud hut than be 'saved' by technology like that. Creepy stuff.

'So you'd leave a fire burning to see if it went out?', she said.

'What? No. I mean ...'

'What if? *What if*? Think if we lost you.'

'Hmmm,' he said.

He had turned down her offer of an all-body scan for his birthday. The ultimate in preventative health-screening, peace of mind, a 3D computerised journey through your own body with a free colonoscopy and take home DVD to share with friends.

'Look at that,' she said, reading the testimonials. 'Just an ache and it was kidney cancer.'

There was a picture of tanned and attractive, smiley doctor and technician types in white coats gazing at computers – and in the background a man's bare legs in black ankle socks sticking from a tube.

'Tumours, cysts, haemorrhage, blockages, heart, bones – your back, for instance – infection. It's quite comprehensive.'

'They put you in a machine,' he said. Cough.

'They're professionals.'

'Doctors? Ha!' Cough.

'Well, rather one machine than a lifetime's worry.'

'I wasn't worried ...' Cough.

PRUDENCE'S HUSBAND would be irradiated by an all-body scan – for the sake of his health. Good idea, she thinks. Nuclear power is a bad idea because of the radiation, she thinks. Some contradiction, perhaps?

He, on the other hand, admires nuke boffins' mastery of nature for the common good, but thinks that doctors who use radiation technologies are quacks. Is he any more consistent?

Yet hearing and telling stories about risk, people like to think they react in a consistent and justifiable way. Do they?

Since exposure to radiation can be measured in standard units, we know how one exposure compares with another. We know roughly how lethal they are, on average, from the good (X-rays) to the bad (radon gas in the home) to a mixture of both (a suntan). This makes radiation a handy test-case, in that people's differing sense of risk about the same level of exposure from different sources is a clue to what personal risk is often about.

What it's often about in the case of radiation is known – no mincing of words here – as 'dread'. For Prudence, at one extreme, the threat of being fried or contaminated by an accident at a nuclear power plant ticks all the fear-factor boxes: it is an invisible hazard, mysterious, ill understood. It seems unnatural. It is associated with those particular nasties cancer and birth defects, provoking concern for future generations with a vague feeling of catastrophic potential. Some feel they can't control or avoid it – it is involuntary.*

We've touched on many of these in previous chapters, but separately. 'Dread' can arise from any one of them, or is sometimes described as a separate emotion altogether, to do with whether people have learned

---

* The media have a field day, even with the very word 'radiation', although heat, light and radio waves are forms of radiation too. So it's important to distinguish between these 'non-ionising' types, which nobody much cares about (except those who feel they are being harmed by mobile phone masts), and the 'ionising' sort, which potentially has sufficient energy to cause changes to atoms and which concerns us here. Ionising radiation can damage cells, which is why radiotherapy is used against cancer, although it is still not clear whether very low doses are harmful. Very high doses can cause acute radiation sickness, while the late effects of radiation exposure can increase the risk of cancers by causing cell mutation.

from familiarity with risk to judge it calmly or with a terrified gut reaction. Either way, dread positively soars when these fear factors mix.

Dread risk is often defined as 'disproportionate', 'excessive' or 'irrational', words that don't try to hide their disdain. If only people weren't so ignorant of the technicalities. And there are differences in reaction to similar radiation risks, no question. But are these attitudes silly, and even dangerous in themselves, or does the notion of dread risk catch something the numbers miss?

When radiation is used to help diagnose or treat medical problems, most people are again like Prudence – relatively untroubled. The history of its medical use has been sublimely blasé. From the first use of X-rays in the 1890s, the magical ability to see inside the body belittled the harm they could cause. Never mind that Edison's assistant Clarence Dally and numerous radiologists died early. After radium was discovered by the Curies, radioactivity was thought of as a source of health and healing. In 1909 Dr E. Skillman Bailey reported to the Southern Homeopathic Medical Association in New Orleans his investigations of Radithor, which enabled him to 'photograph objects through 6 inches of wood'. He used it for medical treatments, although he was no doctor and was said to exhibit 'visible signs of being unstrung'. He even passed a sample around the audience. Nevertheless Radithor – 'Radioactive water, a cure for the living dead' – was a success and sold hundreds of thousands of bottles until forced to close in 1932 owing to bad publicity following the death of the steel millionaire and playboy Eben Byers from radiation poisoning. Byers consumed over 1,400 bottles. The *Wall Street Journal* headline was: 'The radium water worked fine until his jaw came off.'

There were no radiation safety standards until the 1920s. They've grown increasingly stringent ever since: X-ray 'Pedascopes' for fitting children's shoes were banned, as was the 'treatment' with X-rays of children with ringworm, and mental patients with radium. Nevertheless, X-rays and CT scans are still carried out in huge numbers.

People also seem rather unconcerned about natural background radiation arising from sources such as cosmic rays and radon, a radioactive gas from uranium in rocks such as granite. Radon, which can collect in ill-ventilated rooms, is estimated to cause 1,100 preventable lung cancer deaths in the UK each year. But these exposures do not arouse the strong feelings associated with nuclear power.

John Adams cites mobile phones as another example of apparent inconsistency that picks up on one aspect of dread: choice.

> The risk associated with the handsets is either non-existent or very small. The risk associated with the base stations, measured by radiation dose, unless one is up the mast with an ear to the transmitter, is orders of magnitude less. Yet all around the world billions of people are queuing up to take the voluntary handset risk, and almost all the opposition is focused on the base stations, which are seen by objectors as impositions.[1]

Since the phone is useless without the transmitter, that appears odd. Except that you can't choose where the transmitter goes.

In a now classic paper from 1989 on the differences between lay and expert opinion about risk, Paul Slovic said that 'dread' expressed a subtlety that the experts sometimes overlooked. 'There is wisdom as well as error in public attitudes and perceptions. Lay people sometimes lack certain information about hazards. However, their basic conceptualization of risk is much richer than many of the experts and reflects legitimate concerns that are typically omitted from expert risk analysis.'[2]

While acknowledging these concerns, it is still valuable to get a handle on the magnitude of the hazard. So what are the exposures? The units are confusing. It's easiest to focus on the Sievert, which is a measure of the biological effect of being exposed to radiation. One Sievert – or 1Sv – can cause radiation sickness, which might mean hair loss, bloody vomit and stools, weakness, dizziness, headache, fever, skin redness, itching or blistering, infections, poor wound healing and low blood pressure. Not nice.

One Sievert can be broken down into a thousand milli-Sieverts (1 milli-Sievert – or 1mSv – is the US Environmental Protection Agency limit for annual exposure of a member of the public) and a million micro-Sieverts. A tenth of 1 micro-Sievert is about the dose from eating a large banana – Prudence's evening snack – or going through a whole-body scanner at an airport.[3]

The radiation dose chart, Figure 22 (overleaf), is a fun but imperfect way of comparing radiation from the mass of different sources by converting Sieverts into a banana equivalent dose, or BED (imperfect mainly because the radiation in bananas comes from potassium, and the body can regulate this). The virtue of this whimsical unit is that it shows the enormous range of exposures on one scale. Five hundred million bananas – depending, of course, on the size of the banana – is about 50 Sieverts, the equivalent of ten minutes next to Chernobyl power station during meltdown. This helps make the point that many hazards are not a hazard at all if the dose is low enough. After all, who's worried about one banana?

This chart is a very minor publishing scandal, by the way, and may get us into trouble. Most advice on risk communication is to avoid mixing voluntary with involuntary risks. But if one of the complaints about the emotions in dread is that they make people immune to data, then maybe one way to overcome this is to make the data emotionally shocking. Bananas, CT scans and Chernobyl? You bet. Just like horse-riding compared with Ecstasy, it is outrageous, but fascinating, and in our view genuinely helps to put risks in proportion. If perceptions are what often determine risk, then surprising perspectives have a part to play.

Most of our knowledge about the harmful effects of radiation comes from detailed studies of the victims of the bombs dropped on Hiroshima and Nagasaki in 1945. Around 200,000 people died immediately or within a few months, but 87,000 survivors have been followed up for their whole lives. By 1992 over 40,000 had died, although it's estimated that only 690 of those deaths were due to the radiation. Bomb victims had been exposed to radiation roughly according to the dosage shown in the table: the dose declines rapidly, halving every 200 metres, and so, according to a US National Academy of Sciences report, someone around 1½ miles (2.5 km) from the epicentre received around the same dose as from a modern CT scan.[5] What would Prudence say?

The number of victims of the Chernobyl accident is more controversial: a United Nations report says that acute radiation sickness has killed

Figure 22: **Approximate radiation exposure from different sources, converted into: banana-equivalents, equivalent distance from epicentre of Hiroshima explosion, average loss in life-expectancy and cigarette-equivalents**

Exposure	milli-Sieverts	Bananas	Equivalent distance from epicentre of Hiroshima explosion	Average loss in life-expect-ancy	Cigarettes
Ten minutes next to Chernobyl reactor core after explosion and meltdown	50,000	500 million	100 m	50 years	200,000
Dose of radiation that would kill about half of those receiving it in a month	5,000	50 million	700 m	5 years	20,000
Acute radiation effects, including nausea and a reduction in white blood cell count	1,000	10 million	1.1 km	1 year	4,000
Annual exposure limit for nuclear industry employees	20	200,000	2.2 km	7 days	700
Effective dose in worst-affected areas around Fukushima nuclear plant	10–50	100,000–500,000	1.9–2.4 km	3–30 days	300–1500
Whole-body CT scan	10	100,000	2.4 km	3 days	300
Average annual radon dose to people in Cornwall	8	80,000	2.4 km	3 days	300
Chest CT scan	7	70,000	2.5 km	3 days	300
Effective dose in Fukushima prefecture	1– 10	10,000–100,000	2.4–3 km	10 hours –3 days	30–300
A year of normal background dose, 85 per cent of which is from natural sources	2.7	27,000	2.7 km	1 day	100
Mammogram	0.4	4,000	3.2 km	4 hours	16
Nuclear power station worker average annual occupational exposure	0.18	1,800	3.5 km	2 hours	8
Approximate total dose received at Fukushima Town Hall in the two weeks following accident	0.1	1,000	3.6 km	1 hour	4
Flight from London to New York	0.07	700	3.7 km	37 mins	2
Chest X-ray	0.02	200	4.1 km	11 mins	1
135g bag of brazil nuts	0.005	50	4.4 km	3 mins	0.2
Dental X-ray	0.005	50	4.4 km	3 mins	0.2
Eating a banana (or going through airport scanner)	0.0001	1	5.5 km	3 secs	puff
Sleeping with someone	0.00005	0.5	5.7 km	1 sec	Small puff

Note: 1,000 micro-Sieverts (10,000 bananas) = 1 milli-Sievert; 1,000 milli-Sieverts = 1 Sievert.[4]

28 people, and that 6,000 children developed thyroid cancer through drinking contaminated milk, an easily preventable event.[6] Of these, 15 had died by 2005, but the report adds that 'to date, there has been no persuasive evidence of any other health effect in the general population that can be attributed to radiation exposure'.

Others claim the true figures are far higher. It all depends on what you believe about the effects of low levels of radiation. When experts say there is no evidence of extra harm, this is hardly surprising, as even if there had been an increase in cancers downwind of Chernobyl, they would be impossible to detect given the vast numbers that occur anyway. So any estimate would have to be based on a theoretical model of harm.

Using these models, it is estimated that on average roughly one year of life is lost per Sievert exposure.[7] When setting regulations, these effects are extrapolated down to much lower doses for which there is no direct evidence of harm; this is known as the Linear No Threshold (LNT) hypothesis. If we assume LNT, then 1 mSv is 1/1000th of a year lost, which is 9 hours of life or 18 MicroLives. So, as a mammogram is 0.4 mSv, this is equivalent to 8 MicroLives, or around 16 cigarettes. Both average loss in life-expectancy and cigarette equivalent are shown in the table.

The LNT hypothesis is controversial: the people working out the effects of Chernobyl did not use it, and some claim that low doses do not have a proportional effect since the body has time to heal itself. But if we accept the LNT idea, then we can get some remarkable conclusions about the 'nice' use of radiation to help sick people.

For example, a CT scan of 10mSv comes in at 180 MicroLives, or around 360 cigarettes. This may not seem too large for an individual getting a diagnosis, but over large numbers of people it adds up: the US National Cancer Institute estimate that the 75 million CT scans in the US in 2007 alone will eventually cause 29,000 cancers.[8]

None of this was discussed when the apocalyptic visions of destruction brought about by the Japanese earthquake and subsequent tsunami in March 2011 were largely replaced in the media by reports of the struggle to control radiation from the stricken Fukushima nuclear plant. This ticked all the boxes for dread: invisible, uncontrollable, associated with cancer and birth defects. Add an untrusted power company into the mix and the psychological consequences are predictable. The EU Energy Commissioner Günther Oettinger told the EU Parliament that 'There is talk of an apocalypse and I think the word is particularly well chosen.'[9] Was it? From what perspective?

So should Prudence's husband have his CT scan? Would Prudence willingly stand a mile and a half from the Hiroshima bomb?

When people can look at comparable risks of exposure and regard one with utter horror but the other with approval, it might reveal that they are wildly irrational. Or does it rather reveal the limitations of probability as a measure of what danger is really about? Perhaps it simply encourages us to be clear why we really dislike things. To reply, 'well obviously, because of the danger' is a convenient argument, but sometimes not the whole truth.

20

SPACE

W HEN THE BODY of the stowaway hit the street, it sounded
like a slamming door. He was frozen, which, all things considered, was probably for the best. But Prudence didn't think so.

'What if it had hit the conservatory?' she said.

'He, not it,' Norm said. 'Anyway, it didn't.'

'Norm, we sit there for breakfast. Pansy was *there, on her own*, on *that*
chair, when ... my God ...'

'Pru, the chance ...'

'Chance?! What were the chances of a dead body falling?* I mean,
one [she raised her finger] out of the air; two, [another finger] to the
ground, all the way; three, in Basingstoke; four, in the first place? Tell
me that. And what happened? It happened. Well, if that can happen ...
You see. Every morning, we sit in that conservatory having breakfast.
Next thing you know there's a stiff Algerian in your Weetabix. It only
takes one, you know.'

But a few days later it was Norm's turn to look up anxiously.

'You see, asteroids could be it,' he said to Prudence, turning his tea-
cup. It was a lottery, entirely, the chance of the big hit by one among
millions of possible hits in an infinity of cosmic debris.

---

* An Angolan man fell from a plane onto a residential street in Mortlake, London,
in September 2012.[1]

He turned the cup again. It wasn't fear of death that bothered him; more like working out the odds. One day something would fall from the sky, and the chance was both certain and invisible. It made him cross.

'Oh I know,' said Pru, 'I'm the same with pigeons.'

'It's not calculable, the probability, not sensibly.'

'Well it's bloody amazing, the aim.'

'And the damage ... the carnage, potentially.'

'The cleaning, that's for sure.'

'Think of London.'

'Quite.'

'I'm coming round to some form of general-purpose defence.'

'An umbrella?'

'Technically speaking, an umbrella, yes.'

'Always have one.'

'Despite the cost.'

'Trivial by comparison, surely?'

'Absolutely.'

Later, as night fell, he looked up, hands on hips. For a long time he stared, a man alone under the stars, looking for answers. Where was it, the one with our number on it, out in the vastness, the blackness, the unknown? Somewhere, coursing. Except coursing was hares. So not hares. Silent and unseen, towards earth's small plot in this ... this ... fathomless ... this universe, this little O. The path, the uncluttered, no cluttered, the cluttered path through space of the asteroid hurled by a God of limitless thunder that one day would end all human ends was one possible, improbable path ... possibility among countless other possibly ... possible path probabilities, probably. One lump, grain – because it's all relative – in a boundless storm, yes, and the two trajectories of earth and rock predestined by infinite lottery to meet by a cosmic whisker. It was the ultimate doom, reckoning, fate and embodiment of all fear.*

'So,' said Kelvin in the pub later, 'what's it mean for house prices?'

---

* Contemplating human fate while staring at the cosmos has a funny effect on people's prose. See Carl Sagan, who manages proper lyricism in *Pale Blue Dot: A Vision of the Human Future in Space*.[2]

'?'

'Practicalities, Norm, come on.'

> *Email: Prudence to Norm*
> *Subject: Apocalypse*
> Norm, dearest,
> The asteroid was on the news, as you said. Too soon to tell,
> they said, but could be a chance. Wondered if you knew any
> more goss among the star-gazing fraternity, such as whether
> the devastation would be global, as we were thinking of that
> holiday home in Portugal.

'I was just wondering, love, about the asteroid,' said Mrs N a few days
afterwards, 'whether it was worth re-mortgaging – I mean, if we never
had to repay ...?'

NOT LONG AGO Norm discovered he was *l'homme moyen* and hit the
peak of hope and self-belief. Now he faces an existential crisis. Why?
Because with the danger from asteroids he confronts one of the most
absurd averages ever in the ultimate life-or-death calculation. That
makes it not just a distant threat to survival but an immediate threat to
everything Norm stands for. We'll come to this strange average and the
calculation behind it in a minute.

First, who has the more reasonable fear: Norm or Prudence? Heavenly bodies falling from space or human bodies from planes? And why?

Prudence has one advantage: familiarity. She can imagine bodies,
planes and pigeons easier than apocalypse – for obvious reasons. Easier
too, perhaps, to get your head around cosmic doom by thinking in terms
of house prices.

According to news reports, the Heathrow flight path has seen a few
falling bodies over the years, the tragic results of desperation and the
freezing, oxygen-starved atmosphere above 30,000 feet. In 2001 the body
of Mohammed Ayaz, a 21-year-old Pakistani, was found in the car park of
a branch of Homebase in Richmond. Four years earlier, a stowaway fell
from the undercarriage of a plane onto a gasworks near by. No one was hit.

Nobody was at home at the house of the perfectly named Comette family in Paris, either, in the summer of 2011, when an egg-sized meteorite seared into the roof.[3] The rock, blackened as it passed through the Earth's atmosphere, smashed a roof-tile and buried itself in the insulation. It wasn't until Martine Comette noticed the rain coming in and called someone to fix it that she discovered the cause. The rock was thought to be about 4 billion years old, from a belt of asteroids between Mars and Jupiter. When Martine's son Hugo took it to school in a piece of kitchen roll, his friend said it looked like a lump of concrete.

A few months later, in September 2011, a NASA satellite fell somewhere off the west coast of America – prompting concern about how likely it was to land on a human head. At the same time Lars von Trier released his film *Melancholia*, in which 'two sisters find their already strained relationship challenged as a mysterious new planet threatens to collide with the Earth', all to music from Wagner's *Tristan and Isolde*.

All of which gives the impression that there is a lot of heavenly debris about. So what are the risks that a solid object will appear from space and land on your head?

For heavenly bodies the calculation is tricky, partly because of the chance of a truly big hit. And this is what Norm finds so disturbing. An insurer dealing with a car–car collision has abundant direct historical data to help calculate the risks. Astronomers have little. Instead, they devise equations relating the size of an asteroid, how many of them are out there, how often they might hit the earth and what the explosive force would be. These estimates are continually revised and subject to esoteric dispute. The average bottom-line risk that they produce turns out to be truly absurd. We'll come to it in a moment.

There are two main considerations in calculating potential damage. First, the size of the object; second, where it strikes. If a tree falls in a forest and there's nobody there to hear it, does it make a noise? And if an asteroid strikes the atmosphere at a speed of 15 km a second and explodes 10 km above a Siberian forest, where it flattens an area of trees measuring 25 by 25 miles but scarcely anyone is there to see, does it matter? When this happened on 30 June 1908 in Tunguska, the few eyewitnesses willing to talk spoke of a heat so intense it felt as though their clothes were burning, even 40 miles away, of a sky split in half with fire, of being knocked from their feet and running in panic, thinking the end was nigh. It was dramatic, leaving 80 million scorched and flattened trees; otherwise there was some confusion, not many hurt and no direct evidence that anyone perished.

Had the meteorite landed 4 hours and 47 minutes later, it would have hit St Petersburg, or so it has been calculated.[4] One estimate is that such an airburst over New York today would cost $1.19 trillion to insurers in property damage, not to mention roughly 3.2 million fatalities and 3.76 million injuries.[5]

So where an object lands is of at least equal significance to what it is. There is a lot of stuff flying around up there: asteroids made of rock, comets made of ice and frozen gases. At the smallest scale – according to the riveting US National Research Council (NRC) document *Defending Planet Earth: Near-Earth Object Surveys and Hazard Mitigation Strategies*[6] – around 50 to 150 tonnes of 'very small objects', mainly dust, drop to Earth every day. Simply looking into a clear night sky can reveal regular trails of rock or dust burning up in the atmosphere.

Bigger – but only a little more serious – asteroids of 5–10 metres diameter call by around once a year, releasing energy equivalent to around 15,000 tonnes of TNT when they explode in the upper atmosphere, about the same as the Hiroshima bomb. Most go unseen and unrecorded.

Occasionally something gets through and leaves a visible crater or disappears harmlessly into the sea. There is no recent record of human fatality from a meteorite strike, although a few cars in the US have been damaged over the past century and one, the famous Peekskill meteorite car, a Chevrolet Malibu with a boot that looks as if someone took a sledgehammer to it, has toured the world. A cow was also killed on 15 October 1972 in Valera, Venezuela, and duly eaten. Bits of the meteorite were sold to collectors.

A little bigger again, an airburst of a 25-metre asteroid – about the volume of 60 or 70 double-decker buses – would release energy of around 1 million tonnes (1 MT) of TNT, equivalent to around 70 Hiroshima bombs. The Tunguska meteorite is thought to have been about

50 metres across, although some astronomers suggest that Tunguska-like events could be caused by objects as small as 30 metres.[7] Up to this point, even if such an asteroid does strike the earth, there is about a 70 per cent chance that only water will be in the way. So we might still be lucky. Unlucky, and it could kill millions.

Still bigger asteroids begin to fall into the range described as 'continental-scale events', although again it is tricky to say what damage such an impact would cause. On land, it could be devastating. As before, there is a 70 per cent chance it would strike ocean, but this time with more consequence. Models have suggested a 400-metre asteroid could cause a tsunami 200 metres high – almost twice the height of St Paul's Cathedral[8] – although, of course, there is great uncertainty about whether such a wave would break on the continental shelf, whether a population could evacuate and so on.

In the seriously big league, an asteroid more than 1 km across would release around 100,000 megatonnes of energy: equal to about 700,000 Hiroshima bombs and potentially globally catastrophic. Even bigger collisions can and have occurred. More than a cow perished when a 10 km, 100 million MT lump hit the Yucatán peninsula in Mexico 65 million years ago. It left a crater more than 180 km in diameter and probably wiped out the dinosaurs.

That gives some sense of the potential range of damage. The next step is to discover how many of these things are out there and from this derive an estimate of the likelihood that Earth will be in their way.

Fortunately, NASA's Near Earth Object Programme is watching over us, and reports, for example, that object 2009TM8, about 10 metres across, was due to pass closer than the moon on Monday 17 October 2011.

Near Earth Objects (NEOs) are those which could come within one-third of the distance to the sun – around 30 million miles – showing that 'near' is a relative term. When Perry Como sang 'Catch a Falling Star and Put it in Your Pocket' in the 1960s, there were only 60 NEOs known, but by December 2011 over 8,500 had been found and named by the Minor Planet Center. Each year about another 500 are added to the list. There are enough of these for the NRC to estimate an impact equivalent to the devastation from the 50-metre Tunguska meteorite in Siberia about every 2,000 years on average, although if it's true that Tunguska-like events could be caused by objects as small as 30 metres,[9] they would occur ten times more frequently, giving odds of a hit of this size of nearly 50:50 in anyone's lifetime. An even chance that a baby born today will see a Tunguska equivalent, capable of wiping out New York, is about as scary as asteroid statistics can be made to appear. But it is contentious.

Of the big, wipe-out-sized NEOs more than 1 km across, 834 asteroids and 90 comets have been identified, and NASA's Near Object Program estimates that there are only around 70 left lurking out there that we don't know about.[10]

A little closer to home than the 30 million miles that qualifies as 'near Earth' is 20 times the distance to the moon (a total of about 5 million miles), within which range any lump of rock that is found to be at least 150 metres across will earn Potentially Hazardous Asteroid (PHA) status. So far 1,271 PHAs have been found, 151 of which are in the potentially end-of-the Earth as we know-it bracket of more than 1 km diameter.

Still, these quantities are low enough to suggest that a clash with debris of that calibre is only expected every few million years. The NRC stoically points out that 'while this apocalyptic possibility is extraordinarily unlikely to happen in the lifetime of anyone living now, traditional approaches to preparing for disaster would become irrelevant.' Have they not heard of Bruce Willis? Mass extinctioners of the 10 km size that did for the dinosaurs are estimated to come along every 100 million years or so, so not much need to worry there.

These are all averages, of course, seen from a perspective of thousands or millions of years, but the NASA catalogue of NEOs allows us to escape just a little from largely theoretical calculations and talk about what might be in store for those of us living now from specific rocks, all with their own personalised name or number. Each NEO is classified according to its size and probability of impact: these probabilities do not reflect any randomness, since it is assumed that the asteroid is either going to hit us or not, but is simply due to ignorance of its precise trajectory.

The TORINO scale[11] – named after the conference venue in Italy at which it was adopted – expresses the appropriate level of concern:

Level 0 (white) says no problem;
Level 1 (green) is a routine safe fly-by;
Level 2 (yellow) merits attention by astronomers but not the public;
Level 3 means the public should be told;

And so on up to level 10 – a certain collision threatening global civilisation.

At any time there are usually a few level-1s, but Apophis, which is around 200–300 metres across, was given level-2 status in 2004, when it was discovered, and temporarily upgraded to level 4, when a 2.7 per cent chance of impact in 2029 was estimated.[12] New information has shown we're safe after all, although it should be visible to the naked eye on 13 April 2029, when it passes only 18,300 miles from earth. Note that 13 April that year falls on a Friday.

The current catalogue indicates no serious risk from asteroids that we know about. In November 2012 the greatest danger that NASA could point to was a 1-in-500 chance of a collision with the 140-metre 2011-AG5 some time in the 2040s. But most asteroids less than 500 m across remain undiscovered, and although these are unlikely to bring the end of the world, they can take us by surprise: 2008 TC3 was around 2 to 5 metres across and weighed 80 tonnes when it burst above the Sudanese desert on 7 October 2008. It was the first such asteroid to be detected before impact, but there was only 19 hours' warning and one can imagine the crisis if the predicted path had contained a big city. It exploded with a force of around 2,000 tonnes of TNT, and 10 kg of bits were picked up afterwards. Nobody was hurt.

So what is the risk of being killed by one of these rocks from space in your lifetime? This is impossible to assess accurately, and that's where Norm comes unstuck. For although there is rapidly increasing understanding of the chance of an impact, predicting the consequences requires many assumptions (or guesses, depending on your point of view). Nevertheless the NRC report quotes the wonderfully precise figure of 91 deaths expected per year. Of course, this is an average between almost all years with zero deaths and a few events, averaged over millennia, with massive casualties, which once again just goes to show the problem with averages. In fact, that 91 is roughly evenly balanced between the more common small-scale impacts and the very unlikely globally catastrophic impacts. Since there are 7 billion people on earth, this works out at 1/77th of a MicroMort per person per year – the equivalent of about a 3-mile car journey – which comes to the delightfully round number of 1 MicroMort per lifetime from asteroids. Not a lot. And a faintly ridiculous number with no practical purpose. It's not surprising Norm is troubled.

What could be done about an imminent threat? The NRC report identifies four main strategies for mitigation, emphasising the massive uncertainty about the hazards, the technology, and how society would respond. The first, *civil defence*, involves standard disaster management and is suitable for small events, or anything without much warning. If, for example, it was found that Apophis did have a high probability of impact in 2029, the 'risk corridor' could be identified and people warned. The NRC – just like Prudence, Kelvin and Mrs N – identifies possible 'concerns about property values'.

Given a lot of warning and a big budget, space technology might be used to prevent the collision. For bigger asteroids up to 100 m in diameter, with decades of preparation, *slow push* might be used to put the NEO into a different orbit, which is best done by slowing or speeding up rather than sideways nudges. Using a 'gravity tractor' – harnessing the gravitational attraction of an adjacent spacecraft – may be more feasible than actually shoving the rock. Spacecraft have already had close encounters with asteroids: the Hayabusa mission even landed briefly, collected particles and returned to earth.[13]

With decades of warning, asteroids above 100 m and even up to 1 km across might be shifted by *kinetic impact* methods – i.e., ramming with multiple spacecraft. Anything bigger than 1 km would require hundreds of spacecraft to hit it, or alternatively a *nuclear detonation* close by. For 500m asteroids this could be set up in years rather than decades if the political will was behind it.

Anything above 10 km across, the size that wiped out the dinosaurs, is considered essentially unstoppable. Although these apocalyptic scenarios make good, or at least popular, films, the NRC conclude that the main risk is from an unexpected airburst of a small object, less than 50 m, but adds: 'However, as all NEOs have not yet been detected and characterized, it is possible (though very unlikely) that an NEO will "beat the odds" and devastate a city or coastline in the near future.' And there's not much you can do about it.

As for man-made junk, about 5,400 tonnes of rubbish has come down over the past 40 years, and there were 28 re-entries of satellites in 2011. So far nobody has been injured, even when 40 tonnes rained down on the US after the Columbia space shuttle broke up. NASA afterwards estimated that there was a 1-in-4 chance someone would have been hit. And when the remnants of the Upper Atmosphere Research Satellite (UARS) came to Earth in September 2011, NASA said there was a 'one in 3,200 chance of anyone being hit'.

But how are such calculations made? The satellite had been up there for 20 years. It stopped working in 2005 and weighed 5,700 kg, about the size and weight of a double-decker bus. NASA said it would break into 26 objects that would survive re-entry, weighing 532 kg in total, about the weight of eight washing machines. These would be spread over about 300 miles but cover a total damage area of around 22 square metres (around three car-parking spaces), but they had no idea where the debris would land. As one commentator said, you'd think these boffins would have better control of their satellites – it's not rocket science.

The largest object weighed 158 kg, about the weight of an adult gorilla (though that sounds a bit soft – better to think of a couple of washing machines tied together, travelling at 100 m.p.h.). This does not sound encouraging, but the Earth is a big place, with a surface area of 500 million square kilometres (or 500,000,000,000,000 square metres), and so assuming the 22 square metres of bits can land anywhere, there is around a 1 in 23,000,000,000,000 (23 trillion) chance that any particular point will be hit.

So if an individual – let's call him Norm – happens to be occupying that point, minding his own business, then, assuming a random landing place, there is around a 1 in 23,000,000,000,000 chance that Norm will be hit – the same chance as flipping a coin 44 times in a row and coming up heads every time, or slightly better than the chance of winning the National Lottery twice in a row.

But there are about 7,000,000,000 other people on Earth, and so the chance that anybody at all will be hit is 7,000,000,000/23,000,000,000,000 which is one in 3,200, which is just what NASA quoted.*

This chance is low, essentially because people don't cover much of the Earth. It may not seem like that when you are up against a stranger's armpit on the Northern Line, but as anyone taking an intercontinental flight will notice, the globe is covered by an awful lot of not-very-much. If each of us claims 1 square metre, that's 7,000 square kilometres in total, which is only 1/70,000 of the Earth's surface. So if everyone in the world went to the Glastonbury festival, they would only cover Somerset and Wiltshire combined, although you can't even begin to imagine the state of the toilets.

The calculation for falling people would be similar. Let's assume one falling dead body, horizontally occupying say 2 square metres, every seven years, and let's assume an at-risk area about the size of the London borough of Richmond (about 60 square kilometres, with a population of almost 200,000). This gives us a fairly straightforward if crude probability. Norm is comfortable with that. Prudence isn't, it being somehow more real than the end of the world. It works out at about a 1-in-150 chance that one of the 200,000 population will be in the way, every seven years, and, if you happen to live there, a 1-in-30-million chance that it will be you personally, or a 1-in-210-million chance every year.

Is any of this worrying? As ever, it depends on the kind of person you are as much as it depends on the data. In the Lars von Trier film *Melancholia* one of the sisters is alarmed by the prospect of doom, the other is relaxed. Trier was reportedly fascinated by a comment of his therapist that depressed people often remain calm when confronted by threatening or stressful situations – on the grounds that life is awful anyway. The

---

* This analysis assumes that people don't occupy any space. The odds go up a bit if we allow for the width of a body.

German philosopher Schopenhauer made this the basis of a pessimistic view of life in which the only escape from a pointless, eternal failure to gratify human will is through aesthetic contemplation, ideally of music, like Wagner's.

For Norm, who not only isn't miserable enough simply to shrug at existential risk but also feels insufficiently happy for great optimism – he's a middling kinda guy – anxiety about falling objects is more to do with the enormous uncertainty that surrounds the data.

Being average, he should be precisely vulnerable to the average risk of death from an asteroid strike. As he knows, this is the deliciously convenient measure of 1 MicroMort in a lifetime. As he also understands, this is one of the most startling exposures imaginable of the deficiencies of an average, an average that takes in the kind of bolt from the blue – or whatever colour space is – so small that it might dent your car or take out a roof-tile and so rare that the world's media come to take pictures and that it might possibly finish you off if it happens to choose your 1 square metre among the 500,000,000,000,000 square metres on earth, and he combines this probability with the theoretical probability of total wipe-out all round. In other words, it is almost nothing combined with everything.

To produce from that a lifetime risk for any individual of 1 Micro-Mort is arithmetically sound but entirely pointless to everyday life. In other words, this average risk tells us almost nothing, even for *l'homme moyen* Norm. He's not afraid of death, he's afraid for his faith.

21

UNEMPLOYMENT

'I'M SORRY to tell you, Norm, that we're making you redundant.'

'What?'

'We're making you ...'

'Yes, I heard ...'

'... redundant.'

'... what you said. But I mean no. I mean you can't.'

'Norm, you've been a great asset, but ...'

'No, I mean you can't *make* me redundant. I'm either redundant or I'm not, but you can't *make* me redundant.'

'... ?'

'You can't *make* someone unneeded if they're needed, the condition is pre-defined, it can't be imposed, it's ... illogical, completely illogical, it's like saying we're going to make you ...'

'Norm ...'

'... six feet tall. It's basically blatantly against the law, which means it's not even allowed. And you're meant to consult me anyway, and then *if* I were to be redundant because the job's not there any more ...'

'Norm ...'

'... then you can get rid of me, dismiss me, but you can't *make* me redundant. You can't.'

'Norm, we're thinking of giving you the chop because your job's toast and you're scrap. What do you say?'

'Ah. OK ... I see.'

'Good. You're out.'

'Right ... Yes ... Got it ...'

'When?' said Norm.

'How about tomorrow?'

So here he was, coming in to leave. This is what you call a low-probability, high-impact event, he said to himself. In which case, he rather feared he'd miscalculated the probability, as well as the impact.

So. Well. Norm sat at his desk and pulled up his striped socks. He sorted a few papers and put them in the recycling, deleted some emails, made sure one or two people were aware of one or two important bits and bobs and popped in to see what's-her-name in personnel. Norm made a few calls. Norm had a drink at lunch with some of the youngsters, who gave him a nice pen and a card. Norm knocked on the boss's door and said goodbye. 'All the best Norm,' he said. Norm put on his duffle coat. Norm walked past the other desks – 'Cheers Norm' – and dropped his ID at reception. He walked through the revolving door and stood outside on the pavement.

There had been a couple of occasions in life when he had tried feeling truly miserable, but his heart wasn't in it. Perhaps now? The thought briefly cheered him up. Then that night he dreamed of a man in a duffle coat circling the drain. He hadn't expected it, that was all. And so ... erm.

NORM GOT IT WRONG. He thought the worst would never happen. It's the same error some think wrecked the global finance industry in 2008 and contributed to a deep recession – a failure, as Norm puts it, to take high-impact, low probability events seriously. How wrong depends on what happens to him next because, at the extreme, unemployment can kill, as we'll see.

Anyone can get a number wrong, especially if you're trying to put a probability on something that never happened before. And because Norm had never been given the elbow until now, he didn't think too hard about it. It wasn't normal.

When he calls losing his job high-impact, low-probability, he borrows the language of Nassim Nicholas Taleb's book *The Black Swan*: 'I don't particularly care about the usual', Taleb has said.

If you want to get an idea of a friend's temperament, ethics, and personal elegance, you need to look at him under the tests of severe circumstances, not under the regular rosy glow of daily life. Can you assess the danger a criminal poses by examining only what he does on an ordinary day? Can we understand health without considering wild diseases and epidemics? Indeed the normal is often irrelevant. Almost everything in social life is produced by rare but consequential shocks and jumps.[1]

Some events are easier to put out of mind than others. If they're unusual enough or seem improbable enough, or it's hard to work out how they might come about, is there also a temptation to dismiss them? Although it's not as if unemployment doesn't happen to other people. And it's not as if there haven't been financial crashes before. The point is that we get a little too used to what we're used to. And so this isn't really a failure of calculation (as with asteroids in the last chapter). This time it is a failure of imagination. We can't foresee everything that might go wrong, so we take the easy route of assuming it won't. The remedy isn't just better numbers, it's more varied stories that can take us imaginatively out of our comfort zone – one reason why some people are attracted to the idea of scenario planning as a way of thinking what future dangers might be.

Still, the numbers could have helped. In early 2008 in the UK about 1.6 million people were unemployed, a little over 5 per cent of the workforce. Four years later, after a deep recession, more than a million more were on the scrapheap, for an unemployment rate of about 8.5 per cent.

This is the usual way of saying how bad unemployment is, and by implication the risk that you won't have a job: simply look up the current rate. As expected, this risk went up as the economy went down. The effect of recession was that about 3 or 4 more people out of every 100 who wanted a job couldn't find one.

But this way of talking about the risk can be misleading. Over four years there were not 1 million lost jobs. There were more like 15 million, about half of them by people who wanted another one. The number of times that someone was sacked, made redundant, their contract ended or they were otherwise shown the door was many times greater than the number of people counted as unemployed.

This is not because the numbers are fiddled. And it is not because we're counting people who went straight from one job to another. It is because in any economy the job market is a gigantic revolving door through which millions of people pass in both directions, into work and out, spending varying periods on either side. So these 15 million are all people who genuinely had no job for at least long enough to be counted.

When we measure unemployment, we typically measure the number of people who are on the wrong side of the door at any one time. But they are only the net change in the stock of unemployed, the million more who were out of work at that moment, compared with four years earlier. These numbers do not capture the immensity of the flow, or the huge numbers who have felt the chill of being on the wrong side at some time.

This is true not only in recessions. It happens in good times too. Thinking about the great scale of job churn gives a better measure of the numbers of people who experience unemployment, and is a better way of describing what for many is a state of vast and perpetual risk.[2]

The statistics that capture this flow and describe the average person's chance of losing their job are experimental. The Office for National Statistics doesn't track the same people for long enough to be sure that the numbers are accurate. But they'll do, as a rough guide.

They suggest that before the 2008–9 recession, when the economy was by general consent doing OK, the risk of losing a job was between 1 and 1.5 per cent every quarter, much as it had been for years: this is known as the hazard of unemployment, just as 'hazard' is also used for the current risk of death (see Chapter 17, on lifestyle). That is, 1 or 2 people in every 100 with a job would lose it every three months. Though because this is an average, some are more likely to go than others – those with low educational qualifications for example, or those in more casual trades – and some will go more than once.

On top of that, another 1 or 2 in every 100 lose a job to become inactive – unemployed but not looking for work. Sometimes people are classed as inactive not because they don't want a job but because they've given up looking.

Figure 23: **The hazard of unemployment: the proportion of the employed losing their job every three months[3]**

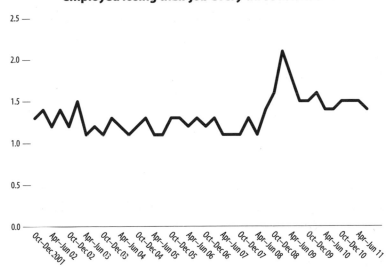

Round up these three-monthly percentages into annual totals and turn them into numbers of real people and we find out where those startling numbers come from: nearly 4 million jobs are actually lost each year to unemployment or inactivity. Four years of that gives the approximately 15 million total of lost jobs from the start of the recession in 2008 to 2012.

There's another surprise in how this risk changes. Before the recession, the average risk of losing a job to become unemployed was about 1.5 per cent. It spiked briefly at about 2 per cent, then fell back. The difference from pre- to post-recession, from steady economic growth to slow or no growth, was equal to only about 0.2 per cent, or 1 extra person in every 500 losing a job every three months.

An additional risk of 0.2 per cent does not sound threatening, but people did feel threatened. Surveys suggested that the fear of redundancy was widespread. In fact, the hazard of unemployment doesn't change as much as we might expect. The hazard of inactivity didn't change much either. In fact, it may even have fallen slightly, meaning a slightly lower risk of becoming inactive.

Figure 24: **The hazard of employment: the percentage of unemployed finding a new job within three months**

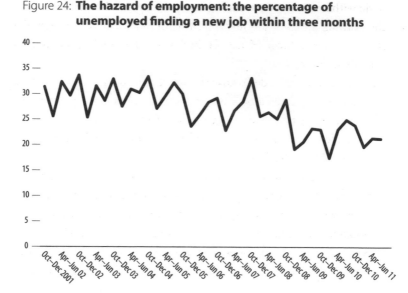

This is a puzzle. Why does unemployment rise so much if the hazard for the average worker doesn't? The answer to the puzzle is partly that we're looking in the wrong place. More significant is what happened at the other end of the story, to the chance of finding a new job if you didn't have one, of getting back through the door once you were out. This is known, curiously, as the 'hazard of employment'.

In the years before the recession up to a third of the unemployed were back in a job inside three months. After the recession, this peaked significantly lower, at about a quarter or less.*

The habit of reporting rising unemployment as if it were all about firms losing business or going bust is misleading; a big part of the story is

---

* The hazard ratios in each direction are proportions of different numbers: that is, the employed are a much larger stock than the unemployed, so a 1 per cent chance of losing a job equals more people than a 1 per cent chance of leaving unemployment. Even so, not finding a job accounts for more of the rise in unemployment. We have also simplified the flows by not going into detail about those around inactivity, which we don't think affect the main point of the analysis.

to do with businesses not opening or expanding. But it is hard to report the nothing of a new job not appearing. It's much easier to show the shutters coming down. Hence 'Becker's Law': 'It's much harder to find a job than to keep one.'[4]

One reason it's harder is that the number of people trying to find work is growing all the time in a rising population. Normally, extra people are soaked up by an increasing number of jobs. This was not often the case between 2008 and 2012. More of the increased workforce stayed on the wrong side of the door.

But it is against the vast, constant labour market churn that the extra threat in hard times of losing your job – affecting about 1 person in every 500, every three months – appears small. To change the metaphor, we are looking at a change in the waves on top of an always heaving ocean.

So the rise in the number of unemployed is the small difference between huge flows of people in both directions, out of work and into work, a difference owed to the difficulty of finding a job as well as the risk of losing one, plus the extra people wanting work every year and lowering the chance for any individual. Amid this great, perpetual flux the *change* in the risk for the average person of losing a job – if they already have one – is small. If you thought it was too high in 2012, you would have to say that it is always too high.

Another way of talking about unemployment is to say that the risk is misery (the probability/consequence distinction rears its head again). According to one study, it is about as bad for our health as smoking. For the young, especially, it leaves a permanent scar on job prospects and wages.

One measure of the consequences is whether unemployment really does become a scrapheap: that is, how long it lasts. In early 2008 about 400,000 had been unemployed for more than 12 months. By early 2012 this was up to about 800,000. So this risk roughly doubled. Your chance of becoming one of them, even if you did lose your job, was still small, but the consequences could be desperate.

That's something only you can judge. Losing a job might be the chance you were looking for to escape and learn dry-stone walling or cash in the redundancy cheque and blow the kids' inheritance on foreign travel, in which case the consequential risk is only that you might enjoy yourself.

Or it might also be catastrophic. In 2010 the TUC, a federation of trade unions in the UK, publicised the story of Christelle Pardo as a warning of the costs of unemployment. With no job, on benefits, pregnant and with a five-month-old baby in her arms, she jumped to her death from the balcony of her sister's flat in Hackney, London.

> Her Jobseeker's Allowance had been stopped because of her pregnancy and this meant that she also lost her Housing Benefit: the local authority was demanding that she return £200 in overpaid HB. She had been turned down for other benefits – her appeals had been turned down twice; her last call to the DWP [Department of Work and Pensions] was made just the day before her suicide. Ms Pardo died almost immediately, her son later that day.[5]

Terrible, but a fair reflection of the risk? Losing a job is unlikely in itself to kill you, but the consequences might. Apart from the loss of income, you are also more likely to divorce, commit crime, suffer ill health and die early. Suicide is often said to be one factor in this rising mortality rate; others are heart disease and alcohol. In short, the unemployed are more likely to be stressed, depressed, broke and sick.

Estimates of the extra risk of an early death vary widely. One study found no effect;[6] others put it at around 20 per cent, with suicide up within a year, cardiovascular mortality rising after about two years and continuing for more than a decade.[7] A third said the overall increase in mortality as a result of unemployment is nearer 60 per cent.[8] That is huge, equal to about an extra 6 MicroLives every day and similar to the extra risk of being an average smoker.

Is the health effect of a day's unemployment really equivalent to that of a dozen cigarettes, bringing you to death that much sooner, as if every 24-hour day without work left you 27 hours older?

But the problem is not as simple as comparing body counts of the employed and unemployed. Untangling cause and effect in these cases is tricky. Are the unhealthy and unhappy more likely to be unemployed anyway, in which case they might die early for the same reason that they don't have a job, but not *because* they don't have a job? Or is unemployment what makes them unhealthy and unhappy?

One study tried to separate the two by ignoring deaths until a few years after unemployment struck, hoping to weed out anyone who lost a job because they were already unhealthy.[9] It also noted that people who are unemployed because they are sick are usually now on sickness benefit and no longer count as unemployed. Its conclusion was that the extra mortality found among the unemployed was overwhelmingly caused directly by unemployment itself.

The size of this fatal effect is still up for grabs, but on current evidence it is probably real. There is also evidence that young people who experience six months' or more unemployment see sharply lower wages in the next few years, and even 20 years later are still earning 8 per cent less than those who stayed in work.[10]

Studies of young people who were unemployed for more than six months in the 1980 and 1990 recessions found that even five years later they spent more than 20 per cent of their time unemployed, and 15 per cent even twelve years later.[11]

High impacts can endure. As Norm has found, risk can take more forms than accidents, and it's not just sausages and cigarettes that can make their effects felt a long way into the future. Nor is it only fatal incidents that can turn out fatally, nor just the known, expected or calculable dangers that are dangerous. So all in all, black swans are even more troublesome than sometimes supposed. First, we cannot predict them; second, we often don't recognise them when they arrive. That is, we might not know whether what we just saw was a black swan until well after it has floated past and its full effect is clear. The 2008 banking crisis was still playing out five years later in 2013, and probably will continue to do so for years to come. Similarly, Norm might not know the risks of unemployment, even after he is shown the door, until years later. So what was the risk?

22

CRIME

PRUDENCE CHECKED HER EMAILS, deleted two financial 'opportunities' from Nigeria, turned off the laptop and popped it under the cupboard in the downstairs loo, took a last look outside and then bolted the front door and put on the chain, switched on the alarm, turned off the lights and went upstairs. She thought again about a dog, especially now that she was alone after her husband had died of prostate cancer, only to run round the same old houses about fleas and mud versus at least a warning bark – and then dismiss the idea. Outside in the wind a tree branch waved into view of the sensor on the security light, switching it on. At the sudden glow in the bathroom window, she twitched, and pulled her dressing-gown tighter. It was two years since they had been burgled. She knew already that she wouldn't sleep.

*K2's diary**
No job. Naturally, go clubbing.
    Eight pints = low dosh.
    Dosh trip with Kate + fresco shag option.
    Old bloke at cashpoint with dog. Dog is snarly dog.
    Dog snarls. Bloke kicks dog. Snarly dog bites bloke hand
    Big cut. Oozes claret.

---

* Kelvin's son.

Also kick dog.

OB says no one kicks dog, kicks yrs truly knee + language.

Kate bends over helps up yrs truly. Cleavage. Nice.

Old bloke gobs at yrs truly but drops dosh.

In consideration of old bloke being knee-kicking/gobbing bastard, and old, shove bloke to ground as he reaches for dosh, grab dosh, tell Kate leg it.

Owing to lack of forethought about efficacy of legging it on bad leg, bloke on floor able to grab bad leg.

With free/good leg kick bloke in head. Language.

Note one foot off ground re kick. Note other foot off ground re grabbed bad leg.

Owing to no feet on ground, fall on bloke.

Dog bites good leg. Bloke bites head.

Pain/claret.

Old bloke fierce bastard. Grinds yrs truly face in ground.

Claret.

Kate multiple smashes OB in head with stiletto.

Claret.

OB stamps on hand, grabs dosh, kicks to various body parts inc. stomach, original bad leg, back. Nicks mobile and wallet inc. cards.

OB legs it + dog. OB nifty for an OB. Conclude later OB probably a crim. Streets not safe.

Kate helps up yrs truly again.

Amazing coincidence ... Emily shows up.

Kate arms round yrs truly while yrs truly cleavage close-up. Hi Em.

Em not Hi. Em grabs Kate stiletto, smashes Kate.

Claret.

Em and Kate wrestle on floor. Looks nice. Have idea.

Em bites Kate. Forget idea. Must help. Kick Em.

Kate says no one kicks a woman, which is tech incorrect as just have.

Kate smashes yrs truly on head with stiletto. Em smashes yrs truly on head with stiletto.

Claret.

Kate, Em leave together. Spit on yrs truly.

Lie in blood/saliva-mix to contemplate strange course of events. In consideration of deterioration in relationship and recent facial disfigurement, decide shag option – fresco/other – off.

Stand up v. slow. Go to pub also v. slow. Note unusual coincidence of two limps. Note v. hard to limp with two limps.

Amazing ... resident pub dodgy geezer offers sale of cash card and mobile. Reluctantly decide to buy back own mobile.

No dosh.

WHICH OF OUR VICTIMS is more typical: Prudence, elderly, alone and anxious, or K2, young, drunk and stupid? Certainly Prudence feels at more risk in her own home than K2 feels, staggering around town late at night. While she's jumpy behind locked doors, he feels indestructible. Reflect on your own impressions of their relative risks and how they're formed, and we'll come to the answer in a moment.

Meanwhile, here are two ways that you might take the measure of crime: 1) examine the numbers; or 2) read crime stories in the newspapers.

The numbers are imperfect and hard to interpret. Stories, on the other hand, are quick and gripping, full of 'psychos', villains, victims and yobs, inner-city no-go areas, trails of blood and vomit from pub to A&E, muggers on trains, sex attackers in shadows, fraudsters cloning cards and conning the old, girl gangs and knife gangs and pushers at the school gates. Asked why they think crime has been rising for the past ten years (when all available data show that it has been flat or tending downwards over the long term), people point to stories in the media.

We need these stories to understand the character of crime. But as with violence against young children in Chapter 3, dramatic events distort our sense of probability. Read or watch these news stories, and the measure of moral decline where none can sleep safely etc. is roughly one vile crime against a pensioner or baby, or a riot. A single instance is about all it takes, and we're going to the dogs.

Although it's not the least bit funny if you're a victim.* About one person in seven says they have a high-level fear of violent crime,[2] like Prudence, and no wonder.† We are highly tuned to scare stories in the news, as news editors well know. Fear sells. We're also highly tuned to the latest information and the latest shock story more than to old news or long-run trends. And fear is useful, up to a point. It is often good for survival. Unobservant and contented never saved anyone from a tiger.

So maybe we are right to grab at every straw in the wind for clues about how frightened we should be, so that we can take care, again like Prudence. For the same reasons we are also alert to local rumour – 'Isn't the one at number 33 some kind of paedo?'‡ – or to a few incidents of the same kind of crime close together – 'all those knife attacks' – and we take special notice of personal experience – anything really, that stands out, any alarm large and small to trigger our attention and help us to a quick judgement.

The psychologist Daniel Kahneman describes the brain as a machine for jumping to conclusions, working according to a law of least effort. This is especially true about fear.[4] If there's something frightening about, it pays not to hang around. Kahneman says this mental habit is a cognitive bias, and crime stories would seem to appeal to it perfectly. Paul Slovic, a colleague of Kahneman's in the 1970s, also showed that vivid events are recalled not merely more vividly but in the belief that

---

* See, for example, the murder of the 94-year-old Emma Winnall.[1]

† A common complaint is that the crime statistics lie. Perhaps people have given up reporting crime because they don't think anything will be done. Perhaps the police fiddle the data. But almost all figures here are based not on what people tell the police or on what the police record. They are separate surveys of what people say has happened to them personally. They are approximate, as surveys always are, but they are reasonable. These surveys used not to include the under-16s, now there are also surveys for 11- to 16-year-olds. One offence not included in victim surveys is homicide, for the good reason that homicide victims can't answer surveys. The homicide data are from what's known as the Homicide Index, maintained by the Home Office.

‡ The mistake of a paediatrician for a paedophile really happened, but not the mob attack sometimes reported.[3]

there are more of them (see also Chapter 4). And crime is often vivid.

The power of the single example over the mass of data is well studied. As Slovic says: 'the identified individual victim, with a face and a name, has no peer.' The same goes for animals, he says. During an outbreak of foot and mouth disease in the UK, millions of cattle were slaughtered to stop the spread. 'The disease waned and animal rights activists demanded an end to further killing. The killings continued until a newspaper photo of a cute 12-day-old calf named Phoenix being targeted for slaughter led the government to change its policy.'

But the plural of 'anecdote' is not 'data', an old statistical saying goes, and the corollary of 'vivid' is not 'likely'. A crime may be devastating for the victim or family, but it's unlikely to reveal that we're going to the dogs, or to say much about the overall risk of becoming a victim. This is almost so obvious as to be trivial, and it is well known. But that doesn't stop a massive online tribute by thousands of people to one dead sparrow, shot for knocking over a line of dominoes in a competition, while the Dutch bird protection agency lamented a lack of interest in saving the whole species.[5]

If we would like a slightly more balanced reaction to single vivid stories, perhaps we should aim to make our description of risks so transparent and convincing that it brings 'immunity to anecdote'. In fact, this has been investigated by psychologists studying people's preferences for treatments, who found that good clear graphics, using arrays of icons as in Chapter 4, could succeed in making people less influenced by stories of wonder cures and ghastly experiences.[6]

Particular events and particular people are usually what stories are about (see our definition in the introduction). In fact, it is often said that without a good dose of the particular, stories lack credibility. That is, authenticity in a story often depends on detail – detail that might apply in this instance and no other. Thus defined, detail is a real-world kick in the shins to generality or abstraction. Detail is the handkerchief that Othello gives to Desdemona and Iago uses to implicate her in infidelity. Small, telling, and 'true to life', detail is the precise story of her murder. It is a claim on believability. The literary critic James Wood has described the importance to fiction of 'this-ness' – or 'individuating form'. By 'thisness', he says, 'I mean the moment when Emma Bovary fondles the satin slippers she danced in weeks before at the Great Ball at La Vaubyessad "the soles of which were yellowed with the wax from the dance floor".[7] Used well, detail is an assertion, then, of realism.

All of which is perhaps the most fundamental way in which stories, both fictional and anecdotal, differ from the abstraction of probability, which fails at detail because your own unique circumstances make the average risks hard to apply to you, or to any other individual.

On the other hand, probability does tell us truths of a different order. In particular, it points out that 'individuating form' can also fail when extrapolated to other people. That's why it was individuating in the first place. Which is why statistics urges us to learn immunity to anecdote.

These two versions of truth speak a different language, but the problem could also be seen as a simple problem of scale. One version, the truth of the story, gains credibility from being personal and particular; the other, probability, depends on doubting the evidence of a single experience, precisely because it is a single experience, until it is aggregated with everyone else's. In each case, what makes one true makes the other a snake. Are they irreconcilable?

For one of the most extreme crime anecdotes imaginable, let's take Dr Harold Shipman. He was what you might call high-risk healthcare, especially if he knocked on your door for a home visit in the early afternoon. Shipman was a serial killer, and as big and vivid a crime story as they come. Reports about named victims filled the media. A handful of chilling cases gave a stark picture of an avuncular-looking man inviting the trust of his patients as he murdered them in their own living-rooms. It was later estimated that he probably killed upwards of 200, mostly elderly, women in good health by injecting them with diamorphine.

But for the rest of us he was not a trend, or indicative of the behaviour of other GPs, or of anything about the general level of crime. Only in Manchester, where all his murders were recorded as having taken place in one year – the year of discovery – did it make much difference to the ordinary ups and downs in the statistics.

Headlines that play on our deeper fears easily mislead. For instance, murderous stranger-danger to our children looms large in public anxieties, but the risk is half that from children's own parents and step-parents. Four out of five adult female victims of murder also know their killer.[8] Alcohol-related violence in England has actually been falling in recent years, contrary to a binge of media coverage, which loves a picture of a lad, off his face, with a bottle. He exists, it is true, but the news is not a balanced sample of behaviour (see also Chapter 25, on media portrayal of danger).

But what if lots of stories appear at once, as when four men were murdered – all stabbed – in separate incidents in London on one day, 10 July 2008? This is no longer one shock headline. Here the story was all about how knife killings were an epidemic. We're not talking anecdotes, we're talking data ... aren't we? Even so, the BBC's Andy Tighe reported: 'Four fatal stabbings in one day could be a statistical freak.'[9] Could it?*

Every murder is an individual crime that can't be foretold. But this very randomness means that the overall pattern of murders is in some ways predictable. This sounds uncanny. It is not (see Chapter 14, on Chance). It is probability doing what it does brilliantly, at the large scale. DS asked the Home Office how many homicides there had been in London in the past year: 170, they said. So he went away and worked out both how many could be expected up to the current date that year, and the likely pattern of murders over three years, assuming they happened at random. On how many days would there be one murder, two, three or four, and how often there would be none at all? The pattern he predicted turned out to fit the real data almost exactly.[10]

He denies consulting a crystal ball. But he does admit to using a little basic probability theory.† He did the mathematical equivalent of

---

* Using 2006–7 as a rough guide, we find that stabbing – or, more correctly, killing by sharp instrument – is the most frequent method of homicide, accounting for 41 per cent of incidents in London, with shooting next, at 17 per cent. We therefore estimate that four stabbings has a probability of $0.41 \times 4 = 0.028$, assuming that the causes of death are independent for multiple murders on the same day, and that the observed rates can be considered as estimates of the risks for each murder, and given that there are four murders in a day.

† In fact, the Poisson distribution, which has nothing to do with fish but is named after a Professeur Poisson.

Figure 25: **Number of days in which there were 0, 1, 2, 3 and 4 homicides in London, 2004–07**

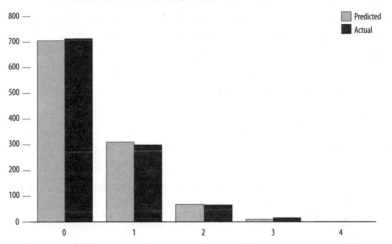

scattering the murders across the calendar like grains of rice, as if they fell randomly. The result is that four murders in one day (four grains of rice on one space) is unusual but not extraordinary, even assuming that the underlying level of violence remains the same. No new or terrifying trend is necessary to produce this result. We would expect it in London around once every three years. We would expect about 705 days with no murders, 310 days with one and about 68 days with two. In fact, the numbers were 713, 299 and 66 – rather close. We can also work out how often we should expect long gaps between murders – a gap of 7 days should occur around 18 times over the 3 years, and it actually occurred 19 times.

It's no surprise that stories can be unreliable as measures of risk. What this work shows is less intuitively obvious: that so can sudden clusters of incidents. One story is not a trend, nor necessarily are four of the same stories on the same day. Clusters are in this respect normal and predictable and will occur simply by chance. The alternative – a perfectly regular pattern of killings with an equal number of deaths every week – is absurd. The Home Office has since adopted the same analysis to help it check on changes in the homicide rate, commenting that 'The occurrence of these apparent "clusters" is not as surprising as one might anticipate.'

On the train home from a meeting at the Home Office at which DS had said we would expect there to have been about 92 murders in London by that point in the year, he picked up a copy of *The London Paper*, which had the headline 'London's Murder Count Reaches 90'.

So we cannot predict individual murders, but we can predict their quantity and their pattern. And by knowing what pattern to expect, we should also be able to spot when something really unusual is happening, when the ups or downs are bigger than chance alone would suggest. When does the pattern of local burglaries look like chance, and when does it look like a new gang on the block? Four murders in London on one day wasn't a surprise, but interestingly four murders all by the same method – stabbing – probably was. Similarly, the pattern of Shipman's killings told us little about the underlying risk of being murdered in the population as a whole, but with better data at the time we might have been able to spot in that pattern the hugely heightened risk that Shipman was a multiple murderer.

Is Figure 26 the most chilling graph ever? It shows the time of day at which people die, and how those in Shipman's care compared with those cared for by other GPs. Shipman's patients had an improbable habit of dying in the middle of his afternoon visiting hours. Patterns are instructive, if we know what to expect.

And while detectives examining one death at a time struggled to decide if that particular death was murder, statisticians could use data to discover both the scale of Shipman's murderousness and to tell us where to look for it. And the genius of their investigation was in the simplicity of a question that seeks out patterns among the patients (and the abnormalities in those patterns) by asking, almost trivially: 'I wonder what time they died?' The 'thisness' of the time of one death in the story of one patient would tell us little in this instance. In aggregate, 'time of death' is revelatory. The 'thisness' of Shipman's behaviour, the particulars of his *modus operandi* – the knock on the door for an afternoon visit to an elderly but often reasonably healthy woman – combined with the numbers for him and for others, enable statisticians to deduce far more

Figure 26: **Percentage of deaths registered as occurring at different times of the day or night for Harold Shipman and comparison GPs in the area**

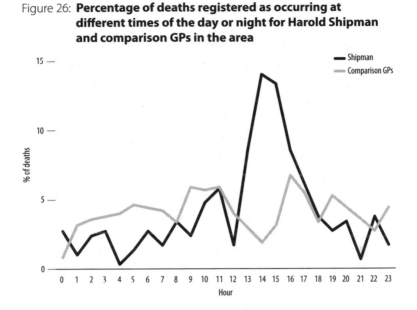

than the detectives could. Figure 26 has a powerful effect on anyone we show it to. The data reveals a truth that particular stories usually miss, because the truth here is contained in the pattern of repetition in many stories, not in just one of them.

Now that we know the need to be cautious about shocking but isolated crime stories, and to understand the kind of patterns that occur by chance so that we also know when a pattern is meaningful, let's take a step further and focus on the overall numbers. If these are up, that must mean something.

Which it does, provided the 'up' has been long enough and big enough. Otherwise, it is liable to the same problem as clusters. The number of crimes goes up and down anyway, by chance, especially in the British Crime Survey, which is based on a sample that will only approximately reflect what's really going on. No new pattern of criminal behaviour, no moral decline or moral revival is necessary for a certain amount of up and down. 'Up' might be chance. 'Down' might be an aberration.

A huge amount of media and political energy goes into short-term

Figure 27: **Average number of victims per 100 people attacked (or average percentage chance of being a victim once or more): England and Wales**

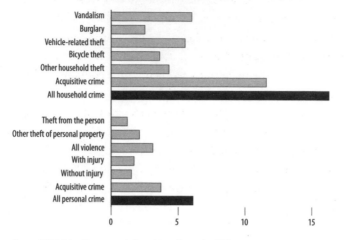

Source: British Crime Survey, quarterly update to September 2011

changes in the crime figures. If discovering trends is what they're after, they are usually wasting their time. Burglaries up 5 per cent this year ... violent crime down 3 per cent and so on. These seldom tell us anything that couldn't be explained by chance or changing fashions of reporting. Burglaries since 2005, for example, have gone up or down every year, but not by enough to say anything other than that they have been broadly flat. Whereas vehicle-related theft has fallen every year, and probably really has fallen quite a lot. Only over the longer term are real changes in crime usually apparent.

Ups and downs also tell us little about the risk of being a victim. 'Up 5 per cent', but from what? There's abundant psychological evidence that people are more sensitive to change than to base levels of risk. Thirty miles an hour can seem fast or slow depending what you were doing a few seconds ago. It's the change that we notice more than the absolute level. As for change in crime, the overall figures for 2011 were as low as they'd been for 30 years. The trends in both recorded crime and the Crime Survey have been down or flat since the mid-1990s.

All this deals with changes in impressions of crime, or changes in the numbers, but it doesn't tell us what those numbers actually are. So finally, what about that underlying, base-level risk of crime?

Using data, a lot of data, rather than anecdote or a cluster of incidents, we can work out the risk of crime simply by taking the number of victims and dividing it by the number of people. And about 3 in 100 people had their home burgled in 2011, about 3 or 4 in 100 suffered violence, but note that the definition of violent crime contains a wide range of offences, from minor assaults such as pushing and shoving that result in no physical harm through to serious incidents of wounding and murder. Around a half of violent incidents identified by both the BCS and police statistics involve no injury to the victim.

Some people experience more than one type of crime, sometimes more than once. Altogether, the chance for the average individual of being a victim of crime in 2010–11 was about one in five.[11] Fear of crime is not the least bit statistically absurd.

But this is still not that useful. It would have been a fair guide to what might once have happened to Norm, being average, but not for you, or for Norm himself now that he is older. By age 65, the risk of violent crime is less than a tenth of that for a man in his early 20s. So the risk for you depends on who you are and which crime you're worried about. For example, women are half as likely as men to be victims of violent crime. If you have been a victim before, for many crimes you're more likely to be a victim again. Ethnicity matters too.

It also matters where you are. Strangely, you were about 16 times more likely to be murdered in Cumbria than in Dyfed Powys in 2010–11, but that turns out to be the cluster problem again. The grim fact underlying the point was that in Cumbria that year the number of homicides went up dramatically when one man, Derrick Bird, shot and killed 12 people. There is always a risk of one-offs, but one-offs do not determine the risk of crime.

A more meaningful geographical pattern is that, if you live in London, you are about twice as likely to be the victim of violent crime as those in Dyfed Powys or Wiltshire. Almost half of the reported robberies in England are in London.[12] Another geographical risk arises with the fact that some types of crime are far more common in poorer areas.

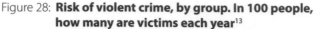

Figure 28: **Risk of violent crime, by group. In 100 people, how many are victims each year**[13]

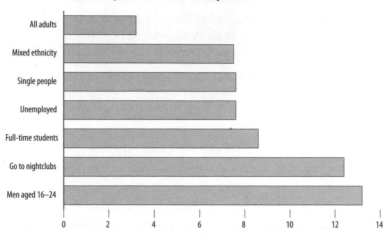

Altogether, a youngish bloke like K2 out and about at night is more representative of crime victims than getting-on-a-bit, middle-class, married Prudence at home in the Basingstoke suburbs in Hampshire. K2's nightclub habit is surprisingly relevant, although part of the risk associated with it is simply a result of his age. All in all, he is a walking risk factor.

For the underlying rate of homicide, if the 172 recorded as committed by Shipman in 2002–3 are excluded, homicide rates peaked in 2001–2, at 15 offences per million people. By 2010–11 the rate had fallen to 12 homicides per million people.

Males – at 16 homicides per million that year – were again more than twice as likely to be victims as females – at 7 per million. That's 7 MicroMorts per year. Compared with other risks, these numbers are tiny, a small fraction of the average 1 MicroMort *per day* for all external causes of mortality. We adopted the MicroMort partly for its small scale, but on this individual scale the average risk of becoming a victim of homicide is about as close to vanishing as a real risk can be. But would Prudence be reassured?

23

SURGERY

T HE PATIENT WAS IN A BAD WAY: white, male, about 85 kg,
history of heart disease, complaining of chest pain and shortness
of breath.

'Blood pressure 85 over 60, falling,' said a nurse. 'Breathing erratic.'

Pulse? Where was the damn pulse?

There was no time. No time for tests, no time to lose. The brilliant
but unorthodox surgeon Kieran Kevlin, 50 years old, silver-haired twin
sibling of a famous former professor at the Sorbonne and at the peak of
his powers, knew he must operate – at once. Haemorrhage? A valve?
His mind raced.

'Can we save him, doctor?' breathed Lara, the surgical nurse, as her
gloved and skilful hands prepared the patient. A wisp of blonde hair
fell across those anxious but still beautiful features beneath the green
surgical mask.

'It'll be touch and go,' said Kieran, looking deep into her compas-
sionate blue eyes, his square jaw set with the steely resolve she had long
secretly loved. 'But for his sake, for his family's sake, for the pride of this
hospital and the values we share, let's give it our best shot.'

From around the operating theatre came murmurs of assent. Kieran
was known as a maverick, but also a medical virtuoso, a man who could
play ducks and drakes with a battery of surgical techniques before breakfast, and follow it with a ten-mile run.

'Thank you, doctor,' said Lara, laying her hand on his gown and sensing the muscular forearm beneath, 'you might never know how much we all admire your... your ...'

'No, no, thank *you*, Lara. I appreciate your always beautiful thoughts. But there's no time for that now. We have a life to save.'

And although he knew the gravity of the crisis, this was also a moment he lived for: the moment of decision and incision, the test of self-belief. To put a knife into human skin, deeply aware of his own fallibility, naturally, and of all that could go wrong by ill-luck or bad judgement, but counterbalanced by training to perfection, even if he did say so himself, which luckily combined in his case with natural genius. He was born to be a saviour of broken bodies, an artist revelling in his craft, a musician transcendent in his calling, primed to wound that tender flesh with a razor-sharp instrument, to make that first cut and see those delicious, delicate red pearls, to rearrange, carve up, butcher, slash and rip and burn and slice like ripe mango, oh yes, and stitch and restore and make whole and *re-create* a God-given human body – this was life and meaning.

He made the cut, swift and deep, and then glanced at Lara to see her soft and yielding blue eyes still upon him. He felt her faith. He could not let her down. Yet he also knew that it would be a damn near thing. He was on the edge, trusting to instinct. Even so, he smiled that roguish, dimpled smile of his, and winked.

Lara's heart filled with desperate joy. If only, she thought, if only he were not so good and devoted to his wife and family. But no sooner had the thought come to her than she felt ashamed – and cursed herself for so selfish, so hurtful a dream. It was not to be. It could not be. She would never know happiness, except for the happiness of being with him now, seeing him in his element, saving lives.

When finally Kieran left the patient to a colleague to clean up, trusting to a smooth recovery, and when his hands met Lara's as they took off their gowns, and they simply stopped, stood and stared into one another's soul for what seemed like eternity, and shortly afterwards somehow fell into the back of his car in a far corner of the car park, where their first love-making was of a sweet intensity that strangely reminded her of his surgery, under the skill of his strong hands, so that the very notion of guilt seemed unkind for an act of such brief perfection – extremely brief, actually – and then when, the very next day, he was suspended and (not long after she realised she was pregnant) later sacked and finally struck off the medical register for negligence following the patient's death for a procedure subsequently condemned as 'ludicrously ill-judged' and, to quote the serious incident report in the ensuing litigation, 'cavalier in its wilful disregard of basic medicine and good sense as if Mr Kieran Kevlin had been more motivated by the daredevil swipe of the scalpel', it seemed to her as if a brave and glorious bubble had popped.

THE DOCTOR–PHILOSOPHER Raymond Tallis says he is privileged beyond the dreams of his ancestors, thanks to modern medicine. That's a bold but fair claim, and Tallis is a superb advocate for the benefits of healthcare.*

Medical drama goes even further. Doctors in fiction are saviours. Sick people are pulled through by techno-miracles or genius. If patients die, the failure is heroic or inevitable. On the whole, our fictional doctors don't kill us slicing out the wrong bits.

Although when a real doctor ordered that an epilepsy drug – phenytoin – be given to Bailey Ratcliffe, a six-year-old boy with a bad fit who seemed unresponsive to other drugs, she got the dose wrong by a factor of about 6. He died. 'I'm sorry,' she said at the inquest in December 2012, 'I made a mistake.'

Doctors do. For all medicine's heroics, and for all the genuine, extraordinary good that it does, white-coat confidence like Kieran's has not always been as well earned as doctors like to believe, or like their patients to think. So here's a question: does popular fiction about life and death risks in medical drama reflect the data? Or by taking on a life and conventions of its own, does it distort popular and professional belief about the dangers?

Kieran is a calculated insult to medicine, partly – wickedly – just

---

* See, for example, his lecture 'Longer, Healthier, Happier? Human Needs, Human Values and Science. Sense About Science', 2007.[1]

to see how it sounds, a parody of the medical hero who kills people by accident and incompetence. It's not the way the story is normally told. We're not suggesting Kieran is the norm, but should the medical story be told that way more often?

Medicine today is coming round to a wider acceptance of error and uncertainty, more willing to acknowledge risks from mistakes and ignorance about what really makes us better, or not. As a result, it is making more of the progress Raymond Tallis describes. But it can still be guilty of what the *British Medical Journal* once satirised as seven alternatives to evidence-based medicine, among them: 'eminence-based medicine', where the more senior the colleague, the less need for anything so crude as evidence that the treatment works; 'vehemence-based medicine', or the substitution of volume for evidence; and 'eloquence-based medicine', for which 'the year-round suntan, carnation in the button hole, silk tie, Armani suit, and tongue should all be equally smooth'.[2]

The growing popularity of statistical hard graft to find out who makes most mistakes, or if people really get better, or worse, because of how we treat them, and how much better, is surprisingly recent. The *Journal of the American Medical Association* announced the arrival of 'Evidence-Based Medicine – A New Approach to Teaching the Practice of Medicine' that 'de-emphasizes intuition [and], unsystematic clinical experience' only in 1992.[3] One critic still maintains that *most* published research findings are false,[4] because studies get it wrong and yet the ones that seem most exciting tend to be those that are published. All of which hardly puts your mind at ease if you're about to go under the knife.

Some medical dramas have picked up this humility. One in particular – the American comedy *Scrubs*, which follows a group of all too imperfect new doctors* – was inspired in part by a compelling account of medical error by a surgeon, Atul Gawande, in his book *Complications*.[5]

Gawande is fascinated by fallibility. His books burst with medical mistakes, and he readily admits his own, including bodging the insertion of a central line into the main blood vessel to a patient's heart. In

---

* In one episode of *Scrubs* the 'hero' is shown on his rounds with a ghost in tow – the ghost of a patient whose death he has caused by medical error.

Gawande's version of the medical narrative, cock-ups, large or small, are routine, even necessary to medical training.

'The stakes are high, the liberties taken tremendous', he writes. 'What you find when you get in close, however – close enough to see the furrowed brows, the doubts and missteps, the failures as well as the successes – is how messy, uncertain, and also surprising medicine turns out to be.'

He describes it as an imperfect science, 'an enterprise of constantly changing knowledge, uncertain information, fallible individuals and at the same time lives on the line'.

But is that the public perception? Or is our mental model dominated by simple stories and ideas of treatment and cure? If the latter, we might underestimate the risks. Hence our story. We wanted to play with the tradition by giving the hero feet of clay, a sack of moral failings and a huge error of judgement.

So, having felt Kieran's disgrace, do you feel any different about medical risks? Probably not. His is just one story. And to be credible stories need – as lawyers say in cases of defamation – a substratum of provable fact. That is, they also need evidence, or at least belief about what the facts and data really say.

As ever, then, some facts and data.

Surgery is simple. A human body is soft, so it takes only a sharp knife to slice out giblets and a saw to hack bits off. The complication is how to stop the patient dying from blood loss, agony, infection etc. Given what we now know about these hazards, it's hard to read about surgery in the past without flinching: at the crudity of the tools, the lack of hygiene and anaesthetic, and, not least, the insane ambition.

Take trepanation, in which part of the skull is removed to reveal the brain, once widely practised either as relief for headache or following injury. The head was particularly prone to damage from slings and clubs and other primitive weapons. The aim of trepanation was to relieve what felt like extreme pressure, release blood and 'evil air', and leave the brain nicely aerated.[6]

Excavations reveal that in Neolithic times as many as one skull in every three has holes drilled or scraped out. The even more remarkable finding is that the original owners of many of these skulls – between

50 and 90 per cent, according to some sources – survived. We know this because the edge of the hole has healed. The procedure was popular in Europe as a treatment for epilepsy and mental illness up to the 18th century, and afterwards for head injury. Cornish miners in the 19th century apparently insisted on having their skulls bored as a precautionary measure after even minor head injuries.

But it was when hospitals took over that holes in the head became especially dangerous. The problem – as with maternal mortality in Chapter 11 – was hygiene, with the infection risk inside a hospital so high that doctors managed to take a mad idea and make it worse. The mortality rate shot up to about 90 per cent. Once again, professionals and institutions were bigger killers even than the procedure itself, taking out perhaps an extra 80 per cent of their patients. Which is why the high numbers of survivors found in 19th century excavations seemed so unbelievable. How could ancient Peruvian natives carry out this operation successfully? Like giving birth in the 19th century, it was far safer to have a hole drilled in your head at home.

Other than having your head excavated with a sharp stone, the only pain relief available for so-called primitives was intoxication. Alcohol, cannabis and opium were the basic anaesthetics until Humphry Davy personally experimented with nitrous oxide, or laughing gas. In 1800 he had the foresight to write: 'As nitrous oxide in its extensive operation appears capable of destroying physical pain, it may probably be used with advantage during surgical operations in which no great effusion of blood takes place.' Naturally, nobody in medicine took any notice for 50 years, while laughing gas and ether were used as party tricks. 'Ether frolics' were hugely popular in the US. At last it dawned on some medical students that the frolickers appeared not to care about injury. Could this be put to practical use, they wondered.

The first public anaesthetic using ether was delivered by William Morton on 16 October 1846 at the Massachusetts General Hospital. The idea soon spread, especially after Queen Victoria grasped at chloroform for the birth of Prince Leopold in 1853, although chloroform later lost favour owing to sudden deaths from heart arrhythmias, now known as 'sudden sniffer's death' among teenage solvent abusers.

To be numbed and put to sleep for an operation is now routine – the World Health Organisation reports that each year there are 230 million major surgical procedures under anaesthesia – with rates strongly dependent on healthcare spending.[7] Anaesthetics are fairly safe now – the UK Royal College of Anaesthetists says that there are life-threatening allergic reactions in less than 1 in 10,000 people, and that most recover.[8] But not all. Around 1 in 100,000 general anaesthetics still leads to the death of the patient. That's a risk of 10 MicroMorts, equivalent to around 70 miles on a motor bike, or around that of a parachute jump. Around half of that risk, 5 MicroMorts, arises from errors made in administering the anaesthetic, which is nice to know. Risks for day-cases are lower, and higher if you are older or it's an emergency operation.

Anaesthetists are fond of saying that the risk is less than that from driving to the surgery, which is only generally true if you come a long way on a motor bike or are a spectacularly reckless driver. If the rates claimed in the UK applied to all the 230 million operations reported by the WHO each year, this would mean 2,300 deaths by anaesthesia – almost certainly a big underestimate.

And hospitals can hurt you in other ways than through surgery, whether it's getting an infection or slipping on a pool of something unpleasant. We can work out roughly the overall risks of a fatal accident during a visit to hospital by taking the 135,000 people who occupy a hospital bed each day in England and noting that, although inevitably some of these die of their illness, not all these deaths are unavoidable. In the year up to June 2009, 3,735 deaths due to lapses in safety were reported to the National Patient Safety Agency, and the true number is suspected to be substantially higher. This is about 10 a day, which means an average risk of around 1 in 14,000, assuming few of these avoidable deaths happen to outpatients. So staying in hospital for a day and a night exposes people, on average, to at least 75 MicroMorts of avoidable death – about the same as giving birth, or a motor-bike trip from London to Edinburgh.

So Kieran's complications story conveys a truth about hospitals: they remain dangerous to our health even while helping it beyond the dreams of our ancestors.

Risk in medicine is unavoidable. But the scope for error adds to it, as does plain bad luck. So it is no surprise that the risk of an operation varies between hospitals and surgeons. The concept of measuring their performance began with Florence Nightingale: after tackling the squalor of military hospitals in the Crimea, she was keen to do the same in England. Obsessed with statistics and a passionate admirer of Quetelet, she viewed the patterns in the data as an indication of God's work. To study them was a spiritual endeavour.

She proposed the collection of 'uniform hospital statistics' to 'enable us to ascertain the relative mortality of different hospitals'.[9] But she was aware that hospitals were adept at fiddling the figures by dumping hopeless cases onto someone else: 'We have known incurable cases discharged from one hospital, to which the deaths ought to have been accounted, and received into another hospital, to die there in a day or two after admission, thereby lowering the mortality rate of the first at the expense of the second.' Nowadays we call this practice 'gaming'. The Victorians could be ruthless in concealing poor performance, just as some hospitals are now, as various scandals suggest, and her grand plan fizzled out.

Forty years after Florence Nightingale, the Boston surgeon Ernest Codman took a different approach to checking the quality of care. Rather than publishing overall statistics, his 'End Results Idea' required hospitals to complete a small card for each patient that explained publicly and in detail whether the treatment was successful and if errors were made. He began this himself from 1900, and even opened his own private hospital in 1911. He claimed that his ideas 'will not be eccentric a few years hence'[10] and, unlike Nightingale, he courted controversy, causing uproar at a public meeting by unveiling a huge cartoon satirising the Boston medical establishment for carrying out expensive and unproven procedures and so grabbing the 'golden eggs' laid by an ostrich representing a gullible public. His scheme, unsurprisingly, did not catch on. He was sacked from Harvard, and his hospital closed in 1918.

There have been modern attempts to emulate Nightingale and Codman, notably for heart surgery, but the quality of the data about hospital performance is probably still not as good as the public thinks it is. Let's take a closer look at one of the exceptions, where the data are not bad, and with a little nosing around we can begin to discover the limits of what we can know about medical risk. This is the coronary artery bypass graft, known as a CABG (pronounced like the green vegetable). CABGs are intended to relieve angina by improving the blood flow to the heart with a piece of artery or vein taken from elsewhere in the body. This type of operation started in the 1960s, and mortality in the US was down to 3.9 per cent in 1990 and to 3.0 per cent in 1999.[11] The UK now reports a '98.4 per cent survival rate', based on 21,248 operations in 2008.[12]

Note the different framing of the information in the US compared with the UK. In the US, people die from surgery, while in the UK they do not survive. This change of framing is a neat device that tends to make performance look better and obscure differences: the difference between two hospitals with 98 and 96 per cent survival, as we would describe them in the UK, looks negligible, while the same comparison expressed as it would be in the US, as 2 per cent versus 4 per cent mortality looks like double the trouble.

Some states in the US mandate mortality reporting. For example, all hospitals in New York State that perform cardiac surgery must file details of their cases with the State Department of Health.[13] In 2008 there were 10,707 CABG operations in 40 hospitals, and 194 patients died either in hospital or within 30 days – a mortality rate of 1.8 per cent. Or, as they would say in the UK, a survival rate of 98.2 per cent.

Surgery on heart valves was a higher risk: of 21,445 operations between 2006 and 2008, 1,120 patients died – a mortality rate of 5.2 per cent, or just over 1 in 20. That's an average of 52,000 MicroMorts per operation, equivalent to around 5,000 parachute jumps, or two RAF bombing missions in the Second World War. Needless to say, this is serious, but presumably the judgement is that the risk without an operation is higher.

This is an example of where we have data, we can define the risks, and that allows us to make comparisons between hospitals. But how useful are they? That sounds like a daft question. If they are real data, what could possibly be wrong with them?

What follows is a trip along the slippery path of working out what the dangers for a group of hospitals really are – even based on relatively good data. Stick with it if you can. It is an object lesson in the difficulty of coming to clear conclusions about risk, and another reminder of medicine's uncertainty about how good it is really.

Taking the numbers at face value, you might prefer the hospital with the lowest rate of deaths, which happens to be Vassar Brothers Medical Center, which had only 8 deaths in 470 operations (1.7 per cent). At the other extreme, we find University Hospital in Stony Brook, which reported 43 deaths in 512 operations (8.4 per cent). But would you be right in this choice?

Maybe Stony Brook was treating more severe patients. It was for this reason that Florence Nightingale decided 150 years ago that crude mortality rates were unreliable for comparisons, since hospitals differ in their 'case-mix'. Ever since there have been attempts to 'risk-adjust' the data to check whether the difference in the number of deaths could be accounted for by the type of patient.

In New York they collect data on the age and severity of the patient's illness and build a statistical equation that tries to say what the chances are that each patient will die in an 'average hospital'. For the type of patients treated in Stony Brook we would have expected 35 deaths, compared with 27 if they were treating only average patients, which shows that Stony Brook really did have a tendency to treat older or more severely ill patients.

But, as we mentioned above, Stony Brook actually had 43 deaths, eight more than expected, so even the case-mix does not seem entirely to explain the high number of patient deaths. With 43 deaths rather than the expected 35, we can say they had 43/35 = 123 per cent of the expected mortality. New York State Department of Health takes this excess risk of 123 per cent and applies it to the overall death rate in New York, 5.2 per cent, to give an overall 'Risk Adjusted Mortality Rate' of 123 per cent of 5.2 per cent which is 6.4 per cent, a figure intended to reflect the risk for an average patient treated in that hospital.

But even a good surgeon might have a run of unexpected bad cases. Kieran's patient dies, but was he just unlucky? Did he have one uncharacteristically off-day, distracted by the lovely Lara? Was Stony Brook

Figure 29: **Heart valve surgery 2006–08: New York state hospitals**

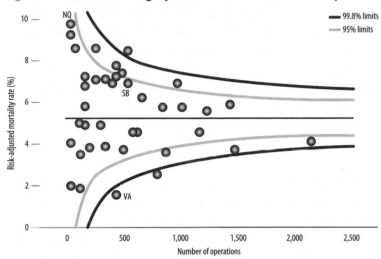

'Funnel plot' comparing the 'risk adjusted mortality rates' for New York hospitals carrying out adult heart surgery. Hospitals carrying out more operations are towards the right. If a hospital lies outside the funnels, there is reason to think their mortality risk is truly different from average. Vassar, Stony Brook and New York Hospital Queens are highlighted.

just unlucky to have a mortality rate (risk-adjusted) of 6.9 per cent, compared with the state wide average of 5.2 per cent? The question is becoming increasingly sophisticated: how do we measure bad luck?

Fortunately statistical methods are good enough to have a stab at it. They depend on making a distinction, mentioned back in Chapter 1 ('The beginning'), between the observed mortality *rate* – the historical proportion who died – and the underlying mortality *risk*, which is the chance a similar future patient has of dying. Rates will not exactly match risks, just as 100 coin flips will rarely end in exactly 50 heads and 50 tails. There is always the element of chance, or luck, or whatever you want to call it.

We can check the role of the ordinary ups and downs of good and bad luck by using a *funnel plot*, which shows the mortality rates for all 40 hospitals plotted against the number of patients treated. Smaller hospitals are to the left and bigger ones to the right. The areas inside the 'funnels' show where we would expect the hospitals to be if the actual risk for each patient was equal to the overall average and the only differences between hospitals were due to chance. The funnels are wider for the smaller hospitals because they treat fewer cases, and so a bit of bad luck can make a bigger difference to their apparent performance. If it were really the case that they were all average and the only difference between them was luck, 95 per cent of hospitals (38 out of 40) should lie in the inner funnel and 99.8 per cent (40 out of 40) should lie in the outer funnel.

In fact, five hospitals lie above and five below the 95 per cent funnel, eight more than we would expect by chance alone; and two even lie below the 99.8 per cent funnel, suggesting they are doing surprisingly well. Stony Brook (SB) lies well inside the funnel, suggesting that its apparent excess mortality could be entirely due to bad luck and there is no reliable evidence that it is an unusually dangerous place to go for treatment. Vassar (VA), on the other hand, has an extraordinarily low number of deaths, even allowing for its case-mix. It looks genuinely good.

The 'worst' hospital is apparently New York Hospital Queens (NQ), with a risk-adjusted mortality rates of 9.5 per cent, almost double the state-wide average. But this is based on only 6 deaths in 93 operations, and so we cannot be confident that this was not just a run of bad luck – it is also inside the funnel.

The UK now produces funnel plots of a Summary Hospital-level Mortality Indicator (SHMI), which compares hospitals according to how many patients die within thirty days of admission, adjusting for the type and severity of their condition.[14] This measure is controversial. For example, what about very sick patients admitted for palliative care only because they are expected to die? Unless allowance is made for these, perhaps hospitals would start using the Nightingale trick of either refusing them admission or rapidly getting rid of patients that were too sick.

One approach has been to allow hospitals to tick a box saying the patients were admitted for palliative care, but the temptation has been to do this for as many as possible as it raises the 'expected' numbers of deaths and so makes performance look better. At the extreme, there is evidence that some hospitals coded as many as 30 per cent of their admissions for palliative care.[15] This box has been removed from the current system.

Measuring hospital risk is a fascinating example of the genius and limitations of statistics. The risks at each institution ought to be calculable – and to some extent they are. But although we can measure differences, and although we can even estimate the extent of luck and bad luck, and though we can use techniques such as the funnel plot to picture all these data and put them in context and show them so that they are easier to understand, we cannot be sure, just as Florence Nightingale could not be sure, that we have eliminated human ingenuity for playing the system. The people factor is still there to mess up the odds. If you want to know the safest place for your operation, check the statistics, but don't expect a simple answer.

24
SCREENING

SHOULD SHE, SHOULDN'T SHE? Prudence was seventy. The invitation – to be screened for breast cancer – lay on the table. It came with a leaflet that told her, more or less, to go; it might save her life. And in the past she had. But now there were these rumours. God, did they know how to make an old woman suffer.

'It's about three to one,' said Norm, offering advice.

'It's well after two,' said Prudence.

'The ratio,' said Norm.

'Oh yes,' said Prudence, 'the ratio.'

'Of the number of people who are treated unnecessarily when the lump is harmless, compared to the number of lives saved.'

'And if it's a positive test, does that mean I'm the one or the three?'

'Can't tell. If we knew that, it would be easy.'

'And when would we know?'

'Never.'

'Never? Even after they'd chopped them off?'

'Even after ... [staring somewhere else] ... so it's for you to decide, really, how to weigh the balance between a life saved and three times as much chance of an over-diagnosis with ... erm ... collateral damage.'

'Yes, Norm. But I'm frightened. Even if I'm old.'

'Oh,' said Norm, looking down. 'Maybe you need to learn to live with uncertainty?' he said, looking up, 'and relax.'

'Relax?'

'Relax.' He smiled.

'I don't want to play the odds, Norm. It's too late. I want not to be afraid. Please, how can I not be afraid?'

'Ah, erm ... erm ...'

FEELING ILL? 'Fine, thanks,' you say. Not worried that the twinge in your chest might be ... something. 'Never felt better,' you say. Ah, then perhaps this well-being hides a risk yet to strike, an unseen killer, lurking in your genes or blood. Are you sure you wouldn't like some professional reassurance?

'Well, now you mention it ...'

Reassurance – peace of mind – is often the health industry message. It's what Prudence wants. And screening sounds like a good way to get it. The impulse to 'find out', to 'check', which Prudence feels powerfully, imagines a day when doubt is put to rest. And it's easy nowadays to find clinics to examine and scan us for a worrying range of diseases that we might have without realising. There are effusive testimonials from smiling people who have been 'saved' by these tests. What could be the harm in having a check up? Possibly, quite a lot.

The story of screening taps into the story of a medical cure that we looked at in the last chapter, and goes like this: Woman (Prudence) is worried. Woman goes for check up/screening. Check up discovers cancer. Cancer is treated. Woman is saved. Screening saves women's lives.

There's a simple linearity here of cause and effect. But is it the only one? Is it, that is, the right story? Prudence is suddenly not so sure, and having her reassurance taken away is hard. She feels this because of recent reports that there is another story about screening – that it also, sometimes, causes harm. The big problem is that we don't know who will be harmed and who will be saved. So in some ways screening creates new kinds of uncertainty and new threats of harm.

How? And how much harm compared with the life-saving benefits?

Consider another screening system: security. Here's something to think about as you inch forward in the queue to be allowed into someone else's country. Suppose you've rounded up a thousand of the usual terrorist suspects, they all declare their innocence, and someone claims to have a lie detector that is 90 per cent accurate. You wire them up, and eventually the machine declares that 108 are probably lying. These are taken away, given orange suits and not seen again for years. No doubt one or two could be innocent, but it serves them right for being in the wrong place at the wrong time.

But then as the years pass and the court cases for false imprisonment mount, you begin to wonder about this 90 per cent accuracy. You go back and check the small print, which says that whether someone is telling the truth or a lie, the machine will correctly classify them in 90 per cent of cases.

But this means, believe it or not, that there were most likely to have been just 10 terrorists in the 1,000, even with '90 per cent accuracy'. The sums are fairly basic: the test would pick up 9 out of 10 of the real terrorists, letting one go free. But there were 990 innocent people, and the test would incorrectly classify 10 per cent of them – that's 99 – as 'terrorists'. That makes 9 + 99 = 108 people sent off to a remote prison, 99 of them wholly innocent – that's 91 per cent of the accused wrongly incarcerated by a '90 per cent accurate' test.

You may think this story is exaggerated, but this is exactly what happens in screening for breast cancer using mammography: only 9 per cent of positive mammograms are truly cancer, and 91 per cent of the apparently 'positive' results – many of which will cause a good deal of anxiety to women, some of whom will go on to have biopsies and other investigations – are false positives.[1] The 10 per cent inaccuracy bedevils an awful lot of healthy people.

Mammography is quite good as screening tests go, as it correctly classifies 90 per cent of cases, but because only around 10 in every 1,000 women screened have breast cancer, most of the positive results are in people who don't have it: false alarms. It's like looking for a needle in a haystack, when a lot of the hay looks like needles, and explains why the vast majority of people who set off airport security alerts are innocent.

Of course, maybe delaying some passengers, giving some anxiety to women and locking up some suspicious-looking people may be a price worth paying. But this is not the only problem with screening tests.[2]

Next, there is the possible harm of the test itself. If we believe the linear no-threshold principle for radiation damage (see Chapter 19), then we can estimate the harm done by imaging healthy people. We've seen that CT scans are estimated to cause thousands of cancers, but most of these will presumably be for some diagnostic purpose. Airport scanning is more controversial, and each 'backscatter' X-ray is limited to 0.1 micro-Sieverts (1 banana). This is only equivalent to a few minutes of background radiation, and is around 1 per cent of the exposure from a five-hour flight itself. A frequent flyer who had 4,000 of these scans would be exposed to radiation equivalent to 1 mammogram, and if we really believe the LNT principle then 100 million frequent flyers would end up with 6 cancers caused by the scanning, as part of the 40 million cancers they would get anyway.[3]

The radiation harms of mammography have been estimated as follows: the NHS Cancer Screening Programme reports that if 14,000 women are screened for 10 years (3 mammograms), around 1 fatal cancer will be caused.[4] If we assume each fatal cancer costs a woman 20 years of life, then that is around 8 MicroLives per mammography, around 16 cigarettes, exactly the figure we arrived at in Chapter 19, when discussing radiation.

A recent US study puts the risk somewhat lower, estimating that if 100,000 women are screened every other year between the ages of 50 and 59, this will lead to 14 cancers and 2 deaths, although the recommended US screening regime would lead to 86 cancers and 11 deaths.[5]

But the main problem with medical screening is what is called overdiagnosis, or harm done by medical care itself.[6] It's a simple idea: things are treated that would never have been a problem anyway.

This is the problem that preoccupies Prudence, recently quantified by an independent report in the UK, which reported that for every woman whose early death from breast cancer was saved by screening, three women were given full treatment for a breast cancer that would never have bothered them if they hadn't gone along to screening.[7]

The curative screening story we told at the beginning was entirely about events, about the bad things that happen and how medicine can save us (see Chapter 4, on non-events). But it is a story that can go

Figure 30: **What might be expected for 250 women who either do or do not attend breast screening every three years between ages 50 and 70, and then following them up until aged 80**

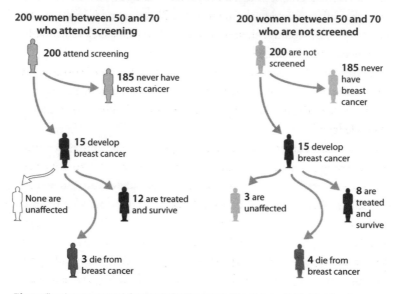

200 women between 50 and 70 who attend screening

200 attend screening

185 never have breast cancer

15 develop breast cancer

None are unaffected

12 are treated and survive

3 die from breast cancer

200 women between 50 and 70 who are not screened

200 are not screened

185 never have breast cancer

15 develop breast cancer

3 are unaffected

8 are treated and survive

4 die from breast cancer

Fifteen will get breast cancer and, if screened, all will be treated, although three will die of their breast cancer. One of the survivors owes her life to the screening, but three other survivors would never have been bothered by their cancer if they had not gone to screening – 'over-treated' cases. Data adapted from Cancer Research UK.[8]

wrong, both in detection of the bad events and by sounding false alarms over non-events, and also even by causing bad events. People who pass through the system follow many different paths and have many different stories, as Figure 30 shows.

So although Prudence wants reassurance, and screening looks like the way to get it, it is in some ways back to the classic problem with risk: it tries to define the individual using probabilities, and these are not certain for you. You will never know which group you are in – whether you benefited from screening or were harmed by it. Screening can help narrow the odds, one way or another, but even relatively accurate screening seldom rules fatal risk out, or in, ever. Norm is right, sadly; we have no choice but to accept uncertainty even if that means more fear.

Prostate cancer is another classic example. The Prostate Specific

Antigen (PSA) test, based on work by Richard Ablin in the 1970s, is used extensively in the US for screening men without symptoms. Ablin now says that PSA testing is 'a profit-driven public health disaster'.[9] Nevertheless, in the UK Andrew Lloyd Webber argued in the House of Lords that 'all men over 50 should have the PSA test and GPs should be encouraged to encourage them to do so'.[10] How can such a remarkable difference in opinion arise?

The problem is that, if you have no symptoms or other reasons to be at increased risk, this simple test can start you on a treatment regime ending in incontinence and impotence, as Lloyd Webber has honestly acknowledged.[11]

It is not difficult to find stories from people who claim their lives were saved by being screened, but it is, however, difficult to say what would have happened if they had *not* been screened. Maybe they would never have been hurt by whatever was found. It is extraordinary how much hidden disease there is, sitting around not causing any problems. When 2,000 healthy people with an average age of 63 had their brains scanned as part of a research project, 145 (7 per cent) had had brain infarcts (strokes) that they had not noticed, while 31 (1.6 per cent) had non cancerous brain tumours. In people over 40 receiving diagnostic ultrasound, 14 per cent of men and 11 per cent of women were found to have gallstones, even though they had no symptoms.[12]

Studies at autopsy for deaths unrelated to any disease – in car accidents, say – reveal the astonishing amount of undetected cancer. In fact, there's around a 50:50 chance that DS has prostate cancer at the moment and MB will have too, very shortly, since 'it is estimated from post-mortem data that around half of all men in their fifties have histological evidence of cancer in the prostate, which rises to 80% by age 80', says Cancer Research UK, although it goes on to point out that 'only 1 in 26 men (3.8%) will die from this disease'.[13]

Unfortunately, screening can't tell the difference between cancers that will turn nasty and those that will sit around minding their own business. Over the last 30 years the US has experienced a large jump in the number of prostate cancers diagnosed, but there has been only a moderate impact on the death rate, even with great improvements in treatment, suggesting the screening has not been of substantial benefit in reducing deaths.[14] But all this activity makes the 'survival' rates look wonderful, as survival is measured from diagnosis, and so the start-time for 'five-year survival' begins earlier, even if people are not living longer. By adding cases that would never have been noticed without screening, the survival statistics appear even better.

Working out the balance between the benefits and harms of screening is notoriously tricky, and the best way is through large trials in which thousands of people are randomly allocated to be offered screening or not. In the US, 80,000 men were divided up in this way, and after 13 years the screened group had 12 per cent more cancers diagnosed but no difference in deaths from prostate cancer.[15] A European study of 182,000 men did find a 21 per cent reduction in prostate-cancer deaths after 11 years in the group offered screening, corresponding to a reduction of 1 death per 1,000 men, so that 1,055 men would need to be offered screening and 37 additional cases treated to prevent 1 death from prostate cancer. But there was no effect on deaths from all causes.[16]

It is very natural for those who have endured diagnosis and treatment for cancer to attribute their survival to the test that first alerted them to their illness. So it is a potentially upsetting message that, of survivors of cancer found at screening, 90 per cent would be alive had they not gone to screening in the first place.[17] People's innate storytelling habit is to connect one event with a preceding one: first this (screening), then that (treatment), leading happily ever after to survival. Not necessarily.

But if not in screening, can we find certainty in our genes? To check this, DS spat into a plastic tube and sent it across to the US, so that a company called *23andMe* could check (screen) some markers in his DNA and tell him whether he could blame his ancestors for his prospects.[18]

But they only provided lots of information about all the horrible things that he *might* develop, including telling him he had an increased lifetime risk of type-2 diabetes, which they did using the graphic in Figure 31. Will he end up being a figure in white or black? Well, he has already lived to 59 and not got diabetes yet, so it's looking hopeful he might be one of the white ones. These are just assessments of the risks

Figure 31: **DS's risk of type-2 diabetes, based on some genetic markers**

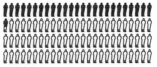

David Spiegelhalter
31.3 out of 100
Men of European ethnicity who share David Spiegelhalter's genotype will develop type-2 diabetes between the ages of 20 and 79.

Average
25.7 out of 100
Men of European ethnicity will develop type-2 diabetes between the ages of 20 and 79.

using a few items of information abut his genetic make-up that could have been obtained when DS was a baby. Now things have changed, and so the odds have too.

To find out whether he had a genetic risk factor for Alzheimer's disease, DS had to tick a box agreeing that he really, *really* wanted to know this information before it would be unlocked. Which he did. And the answer was? He isn't telling.

So is this the future? And how much anxiety, investigation and unnecessary medical care will it bring? Because if it's a future that chases reassurance, it might in a vast number of cases simply remind us, perhaps painfully, how much we will never know.

25

MONEY

IN THE DREAM Norm was by the sea again with his little brother, digging a hole in the sand just like they did on that endless beach at Skegness. He liked it outside, liked the sea and the air. It was the time Dad bought them those red metal spades with wooden shafts and handles. The spades were ace, they said, ace. He gave them money for ice creams too. Norm felt the coins in the pocket of his shorts. He liked that: the roundness, the promise.

The sand was firm and just wet enough to stay up as they carved the sides of the hole, steep and square. Their arms dug, turned and tipped. They liked how the sand gave way to the spade, how it cut in slabs. Then Norm's brother began another hole. Now there were two holes with a wall between them two feet wide. They cut foot holes in the sides to climb out.

'A tunnel,' said the little brother.

'Yeah, a tunnel,' said Norm, 'sort of.'

Carving slices of sand, they joined their two holes with a tunnel. They widened the tunnel and widened it some more, slice by-slice, with their red spades, squatting and cutting, watching for cracks in case it was going to fall. And when it was done they were glad and had two holes nearly as deep as the boys were tall. There was a bridge of solid sand, the tunnel perhaps two feet high.

'Crikey,' said Dad, who came over with trousers rolled up, 'can you get through?'

Norm's brother jumped in with a sandy thud, twisted onto his hands and knees. He was little. He squeezed through the tunnel and up the other side.

'Easy,' he said.

'You'll be all right,' said Dad to Norm. 'I'll watch.'

Norm jumped down. It was grey and a bit wet at the bottom of the hole. He hadn't noticed before, when he was digging. He peered through to the other side and saw that it was grey and wet there too. He knelt down. The tunnel was dark and the roof low. He wanted to be quick like his brother was quick, and he dipped his head and shoulders and crawled, but his bum caught the bridge and for a moment he was stuck and then he backed up fast.

'Can't do it.'

'Go lower.'

Lower would be not flat, but not strong and quick like it was on hands and knees. It would be in between, and that was awkward and bad. It was bad to move his face and stomach closer to the sand, with so much sand on top. He dipped down, lower, and crawled, and bumped, and so he went lower again, deep in the smell of wet sand, further from the light, his hips low and his arms bent, his lips gritty and his face close.

And then his head was through. But his body was still in, with the thick block of sand over him, and he was staring straight at the sand wall of the side of the hole and the turn on this side was awkward, and he bumped the walls, bumped the roof, and he was stuck again and needed to twist, and puffing, something tight, he … around … that way … and was out, and up … and felt something go inside, and cried. So they jumped on it until it fell. Then he felt in his pocket, and it was empty.

Norm woke up. He put on his dressing-gown and went downstairs to check the balance in his savings account and look again at his pension statement. Silly. He knew what it said. Silly, Norm, to feel so put out, so vulnerable, again, still.

AS IMAGES GO, penniless and in a hole isn't subtle. But then nor is the kind of fear that takes you, unreasoning, by the throat, or the strange way that associations form. Nightmares and phobias don't go away because we tell ourselves not to be silly.

Norm is old. He's seen some life. He has experience. But his anxiety here goes back a long way, and for him no amount of wisdom or calculation has cured it. This is the doctrine of the searing memory: the deep personal scar on every judgement. When you can still taste the panic in your mouth like sand, what chance the objective calculation of abstract risk?

For Norm, who trusts in logic, this hurts: once again, his mind won't obey his own orders, all because of one tyrannical moment years ago. Out of proportion? It makes no odds. For decades he has been telling himself to grow up and be reasonable, but little by little Norm has also been learning what it is to be human.

Phobia is an extreme example of the availability bias that we met in Chapter 4 (on the problem of framing). Availability bias, as we saw, means whatever comes easiest to mind. Everyone is affected by availability bias, although we've afflicted poor Norm more than most by giving him a phobia. It sticks in his mind. It comes to mind often. He can't help it. Sorry, Norm.

Daniel Kahneman argues that there is evidence people can fight availability biases if they are encouraged to 'think like a statistician' and work out what it is that might be shaping their opinions. You can do this, he says, by asking questions such as: 'is our belief that thefts by teenagers are a major problem due to a few recent instances in our neighbourhood?' Or 'could it be that I feel no need to get a flu shot because none of my acquaintances got flu last year?'

The phobia haunts him now because he feels his last years are especially vulnerable, afraid of being penniless and powerless. It's not much of a sunset, but is it true? Or, if he's typical, will Norm burn the kids' inheritance on Saga cruises, flush with an index-linked pension and fat on housing equity withdrawal?

What's certainly true is that people are living longer (see more on this in Chapter 26). That should be cause for celebration, except that so many seem to fear they won't be able to manage, perhaps even that they face misery and poverty. To others, the generations now reaching retirement have had it all – and seem hell-bent on blowing every penny. Whatever the truth, instead of the blessing of a long life, we talk of the risks and burdens of age – either the burden of enduring it or the burden of paying through the nose so that others can soak up the sun.

So who is right, and which image of retirement is true: the risky one or the indulgent one?

Both, to some extent. There are pensioners who have done well, and there are plenty who have not. Since this chapter is about financial risk and insecurity in retirement and old age, we will concentrate mostly on the have-nots.

Old age historically has not been a time of plenty. It is estimated that in 1900 about 5 per cent of the elderly overall and about 30 per cent of the over-70s were in the workhouse, still operating under the Poor Law of 1834.[1] Lives inside were often harsh by design, to deter the able-bodied, although workhouses did at least provide medical care.

But they were harsh for the infirm too, and it's less well known that they became a destination for the old. Still more elderly people relied on payments known as 'outdoor relief' – small supplements to scraps of work, charity or family help, so that with luck they could remain outside the walls of the 'pauper bastilles'.

Most elderly people relied on others to some extent. Many were well cared for, but a big minority were not, and 'the extreme smallness of their means' was noted by social reformers like Charles Booth and See-bohm Rowntree.[2] Only after the appointment of a royal commission in 1905 was it suggested that deterrent workhouses be reserved for 'incorrigibles such as drunkards, idlers and tramps'.[3] George Orwell wrote of standing in the queue for 'the Spike' with a tramp, 'a doubled-up, toothless mummy of seventy-five'.[4]

With the development of pensions in the 20th century, retired life steadily improved, although women were still disadvantaged, often receiving pensions only through their status as wives or widows. Even so, most commentators agree that retired life became far less grim – with 'gold-plated' pensions and voluntary early retirement at the top. Recently, some have argued that it's about to get worse again and that the next generation of pensioners faces a meaner future.

Figure 32: **Percentage of pensioner households living below 60 per cent of the median household income, after housing costs ('living in poverty')[5]**

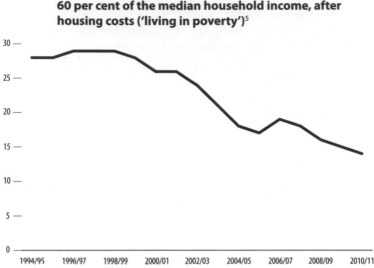

There are two big fears here: first, not having enough to get by for so long as you can be independent; second, being cleaned out by the cost of care.

So, to the data. About one in every six pensioner households still meets the standard definition of living in poverty – an income below 60 per cent of the median – but this has fallen sharply, by about 45 per cent since 1999. Retirement and old age in relative poverty are still a fact of life for nearly 2 million people in the UK, but it has improved rapidly.

And other groups fare worse. Current pensioners have reached the surprising point where, as a group, they are less likely to be poor, on average, than almost any other section of the population, with almost the only exception being younger people working full-time.*

---

* The standard way of measuring income here is by household. Since different households have different numbers of people in them, the figures are adjusted to take account of this. So it is assumed that a single person living alone needs 67 per cent of the income of a couple to achieve the same standard of living. Similarly, people with children need more than a couple. By this means, all households can be compared on the same scale, approximately.

Figure 33: **Percentage of different types of household 'in poverty'**[6]

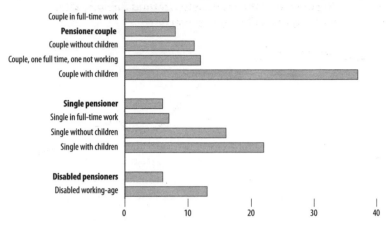

Source: Households Below Average Income (HBAI) 1994/95–2010/11

In fact, the very poorest pensioners do better after retirement than before. The English Longitudinal Study of Ageing has followed thousands of people as they grow older. The latest report says that 'for people with low incomes (less than £150 per week) before retirement, income tends to increase on entering retirement, perhaps as a result of state support for pensioners on low incomes'.[7]

So the image of the impoverished pensioner is misleading. Not because some aren't struggling – they are – but because this is not a characteristic of the way we treat the old. It is a characteristic of the poor – of all ages – and it tends to be less true of the old poor than the younger poor. This makes an important point: most people who are poor in retirement were poor before. Retirement probably didn't cause it.

For those higher up the income scale, retirement is unlikely to mean poverty, but it's often a bigger jolt. Incomes here do fall, typically by about a quarter. Net weekly income in about 2009, after taxes and benefits, dropped from an average (Norm again) of a bit less than £400 to just under £300.

Another way of measuring how well off people are is to look at their spending rather than their income. And here huge rises in fuel prices, for example, have wiped out a chunk of the increased income of the past few years, just as inflation has for other age groups. But since pensioners tend to spend more on basics, such as heating and food, and basics have tended to rise in price faster than inflation generally, pensioners have suffered more from rising prices than others, so the relative improvement in their financial position over time is not as good as it looks.

Even so, overall and in the long term, life in retirement has improved dramatically, especially for the worst-off. Pensioners have become steadily less likely to be classed as poor. For the better-off, retirement does mean a drop in income, but not normally a desperate one.

Yet one cost can change all that. Because although you can insure yourself against unemployment or illness, you can insure against your house burning down or accidents at home, on the road, or abroad, you can insure your pets and in financial markets you can even buy end-of-the-world insurance, you cannot insure against the potential costs of care in old age.

In 2012 this was just about the only potentially expensive lifetime risk that could wipe you out with no way to stop it. There was simply a messy and often inconsistent system that punished the unlucky with the loss of all their assets, including the house. Even at that price, the quality of care varied hugely.

If you can't insure against a risk, you have no choice but to take the cost on the chin if the worst happens. That can screw up life for everyone, even those who turn out not to need care, because the problem is that you never know if it will be you.

It has been estimated that a quarter of people aged 65 will need to spend very little on care over the rest of their lives. Half can expect care costs of up to £20,000, but 1 in 10 can expect costs of over £100,000. Some could spend hundreds of thousands of pounds, and there is no knowing who, no way of predicting in advance what the costs might be for any one of us.[8] For those who have little or nothing, the state will pay. For the rest, the ruinous tail-end risk frightens them and frightens insurers out of the market altogether.

Uncertainty about who will be in that tail of high need might help explain a puzzle about the behaviour of elderly people: that at a time when incomes fall for most, and economists expect them to begin to use their savings to help keep up their standard of living, many are still saving hard, perhaps because they fear they will be in the unlucky half. If even that proves optimistic and they are among the unlucky 1 in 10 with more serious needs, saving at this stage of life probably won't make any meaningful difference. For this reason if no other, Norm is right: life after retirement is still a financial lottery in which some can lose almost everything.

THE END

THE ASTEROID MISSED. The odds that it would strike earth had lengthened, as they tend to do following initial discovery, once the trajectory is refined. Even with years to go, it was clear the Apocalypse was postponed, again.

Stargazers watched the fly-by in fascination with a quiet thrill at the thought that this was the one that could have been. 'What if ...,' they said.

Norm stayed inside. And S043 passed quietly about half-way between the Earth's atmosphere and the moon, then sailed into oblivion. It had all been a false positive. A non-event. House prices rose slightly.

In the evening he stood straight and proud in his pyjamas, a stub of Aquafresh on the toothbrush that hung by his side. He noticed a slouching Norm reflected in a dark window as if his appearance was news to him. He didn't bother with all that any more.

A minute passed. The stub of toothpaste began to peel off. The pipes of the central heating warmed and stretched.

'Any pike out there?' Norm asked out loud, at last.

No answer. He'd asked before but was less sure nowadays that he wanted one. Did it matter – the answer? What if someone told him, and then told him what to do about it, laid out all the answers? Norm sighed. No need for choices if all the answers are given.

He smiled. Then he looked at Norm with narrowed eyes.

'I see through you,' he said.

Paler and paler these days, it was almost true. Growing old lent his skin a soft transparency that was not white, nor grey, nor greyish, nor reddish, nor brownish. Tupperware skin, Mrs N said. Tupperware hair too. People ignored him. They said they couldn't make out what he was saying.

'Everything all right?' she called from the other room.

'Fine,' he said.

Norm held his own gaze, fixed as a Rembrandt. All his life he had tried to reduce fear to probability. Strangely, he felt no fear now.

'Probability ...? Doesn't exist,' he said, just to be rude to the man in the window, who was rude back, sharing one feeble voice. He stared harder, his eyesight so poor these days he could hardly make out ...

'The average of what?' he mumbled ... what with the poor light, feeling almost weightless himself – as the toothbrush dropped.

No pike, then, not for him. He reached down with one hand like the boy who stood – was it on the bank, or was it a dream? – limbs frail and light, and lighter, as he reached, or thought he reached, and felt nothing. Curiously, Norm had disappeared.

Mrs N woke the next morning and rolled over. As she made tea downstairs, Kelvin called about wanting a lift. 'Have you seen Norm?' she asked.

'Who?' said Kelvin. He was old. But perhaps he was right.

'Never mind.'

Later, Mrs N noticed his pyjamas, crumpled in the bathroom and a toothbrush on the floor.

Later that day, coming out of the bookies in his wheelchair, Kelvin had a heart attack – all those fags and burgers, probably – and went out suddenly. It was not abnormal for a man of his age and habits, which would have irritated him no end. There's an average even for unusual people: the unusual-people average, and Kelvin, unknowingly, was typically unusual.

Years later, Prudence had her last coherent thought about Norm as she dribbled into her All-Bran and hoped that, wherever he was, he was safe, but she felt horribly confused by the strangeness of it all and it wasn't long before she had forgotten him again. Having looked after herself so well for so long, her body lasted until her mind gave up. She drifted on for five more years or so, happily unaware at last, loved by her family until the end.

On the basis that at least one unusual thing happens to everyone on average, even to the average, it could be argued that Norm's strange disappearance and presumed death were unsurprising. Except that for some, the unusual thing that happens is that nothing much happens at all.

DID NORM HAVE TO DISAPPEAR? We'll answer that in the next, and last, chapter. Though they all had to go somehow. In that respect, death isn't a risk at all. It has a probability of 1. All that's risky about it is that the reaper turns up sooner than you think right or proper. For some of us, the right moment can never arrive. For Kelvin death seemed like one more rule, so of course he'd have wanted to rebel. For Pru, when the risk really was creeping right up behind her, just for a change she couldn't care less. But the timing was, for all of them, in its way normal.

And normal is ...?

The 90th Psalm in the King James version of the Bible declares that 'the days of our years are three-score years and ten', although up to recently you had to be fairly lucky to reach this use-by date. No English monarch survived until 70 until George III, in 1820, although they did have to cope with draughty castles, bad sanitation and occasional violence. Some historical figures managed it: Augustus Caesar conquered 75, Michelangelo hammered on to the amazing age of 88.

The number 70 was important for another reason: since the ancient Greeks there had been a superstition about the risks of ages that were multiples of 7. In particular, the 'climacteric years' of 49 and 63 were thought positively dangerous.

Partly in order to combat this belief, in 1689 a priest in Breslau in Silesia (now Wroclaw in Poland) collected the ages at which people died. The data eventually found their way to Edmond Halley in England, who took time off from discovering comets to construct the first serious life-table in 1693, which uses estimates of the annual risk of dying to work out the chances of living to any age.

Figure 34: **Halley's original calculations for what we would expect for 1,000 people starting their first year**[1]

Age. Curt.	Per-fons	Age. Curt.	Per-fons	Age. Curt.	Per-fons	Age. Curt.	Per-fons	Age. Curt.	Per-fons	Age. Curt.	Per-fons
1	1000	8	680	15	628	22	585	29	539	36	481
2	855	9	670	16	622	23	579	30	531	37	472
3	798	10	651	17	616	24	573	31	523	38	463
4	760	11	653	18	610	25	567	32	515	39	454
5	732	12	646	19	604	26	560	33	507	40	445
6	710	13	640	20	598	27	553	34	495	41	456
7	692	14	634	21	592	28	546	35	490	42	427

Age Curt.	Per-fons	Age. Curt.	Per-fons	Age. Curt.	Per-fons	Age Curt.	Per-fons	Age. Curt.	Per-fons	Age. Curt.	Per-fons
43	417	50	343	57	272	64	202	71	131	78	58
44	407	51	335	58	262	65	192	72	120	79	49
45	397	52	324	59	252	65	182	73	109	80	41
46	387	53	313	60	242	67	172	74	58	81	34
47	377	54	302	61	232	68	162	75	88	82	28
48	367	55	292	62	222	69	152	76	78	83	23
49	357	56	282	63	212	70	142	77	68	84	20

Age.	Perfons.
7	5547
14	4584
21	4270
28	3964
35	3604
42	3178
49	2709
55	2194
63	1594
70	1204
77	692
84	253
100	107
34	
	34000 Sum Total.

By year 15 there would be only 628 left alive, of whom 6 would die before 16, a force-of-mortality of 6/628 = 1 per cent. By year 75 there would only be 88 alive, of whom 10 would die before 76, a force-of-mortality of 10/88 = 11 per cent.

He found no evidence of any increased risk at age 49 or 63, so neatly demolished the idea of climacteric years. His comet also duly turned up, as predicted, in 1758, when Halley would have been 101 if the force of mortality had not finally got the better of him. Halley's life tables only went up to 84: he estimated there was a 2 per cent chance of reaching this age, and just to prove it he died when he was 85, thus by some margin outliving Bill Haley, also of Comet fame, who rocked around the clock only until 55.

'Force of mortality' was, in fact, a technical term for the annual risk of death, now known as the 'hazard'. The rough shape of the hazard curve – as shown in Figure 35 – seems to be a constant throughout history, and reflects the ages of man: a high hazard just after birth declines to a point of greatest safety and then starts increasing again, although the precise pattern in childhood and young adulthood depends on the contemporary state of infectious diseases, war and, nowadays, recklessness with drink, drugs and driving.

Figure 35: **The annual risk of death – the 'force of mortality' – derived from Halley's data for the 1680s**

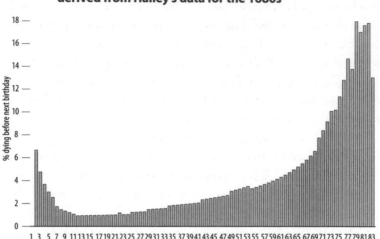

About 1 per cent of 15-year-olds are estimated to die each year, and around 11 per cent of 75-year-olds. For UK men in 2010, the corresponding annual risks are around 0.02 per cent and 3.5 per cent.

In 1825 Benjamin Gompertz, barred from university for being Jewish, formulated his 'law of mortality', which says that after the mid-20s the annual risk of dying increases at a constant rate. This is extraordinarily accurate, and every year between the ages of 25 and 80 the risk of dying has a relative increase of around 9 per cent, as we already mentioned in Chapter 17. Gompertz did his best to fight against his own law but finally died when he was 86.

Longevity is usually summarised by life-expectancy – the average length of life – but averages can be misleading, as we've seen so often. Life-expectancy is strongly influenced if there are a lot of deaths in childhood, in which case the survivors may live to a good age, but the average will still be low.

In 1958, when he wrote 'When I'm 64', Paul McCartney was only 16, an age when 64 seems old – and particularly in an era when people thought retirement meant putting on your slippers and waiting for death. But even the chance of surviving childhood and reaching 16 has changed dramatically. For example, if we consult that wonderful resource the Human Mortality Database, we find that back in 1841, a frightening 31 per cent of children born in England and Wales died before they were 16.[2] But if you survived, there was nearly a 50 per cent chance of reaching 64. By the time the Beatles recorded 'When I'm 64' for *Sgt. Pepper's Lonely Hearts Club Band* in 1966, only 2.5 per cent of children died before 16, and surviving girls had an 85 per cent chance of reaching 64: for boys it was 74 per cent, the difference partly reflecting the unhealthy lifestyles of so many men.

By 2009 less than 1 per cent of children died before 16, and after that the chance of reaching 64 had risen to 92 per cent for women and 87 per cent for men, giving an overall life-expectancy of 82 for women and 78 for men. This does depend crucially on where you live, which is often a function of where you can afford to live: if you are fortunate enough to be a resident of Kensington and Chelsea, then a mostly well-heeled lifetime (though there are pockets of poverty even in K&C) of tea and crumpets, or more likely these days gazpacho and twice-cooked pork belly with an onion & apple velouté, is expected to last until 90 for females and 85 for males, whereas in Glasgow City rather different consumption and material lifestyle is associated with 12 fewer years for females and 13 fewer for males.[3] Although it's a measure of the speed at which people are putting on extra years of longevity that today's disadvantaged Glaswegian enjoys the same expected survival as the average man in England did in 1983.

Life-expectancy in the UK has been steadily going up by 3 months a year for decades. That is as if, after using up 48 MicroLives just slogging through a day, you are given 12 back again by those nice people who build drains, give us injections, stop us smoking, sell us low-fat milk and treat us in hospital.

But this is nothing to the extraordinary changes that have occurred elsewhere. In 1970 people in Vietnam had a life-expectancy of 48. It is 75 now, a transition that took England and Wales more than twice as long, from 1894 to 1986.

Behind the cold columns of numbers in historical tables lie powerful events: apart from the world wars, Napoleon's march on Moscow (which killed 400,000 men) temporarily reduced life-expectancy to

23, the influenza epidemics of 1918–19 took 10 years off life-expectancy for females in France,[4] while AIDS has meant life-expectancy in South Africa declined from 63 in 1990 to 54 in 2010.[5] In 1901 life-expectancy for Black males in the US was 32, with 43 per cent dying before 20, while for White males the corresponding figures were 48 and 24 per cent.[6] After 100 years the gap in life-expectancy was still there, although it had narrowed from 16 years to 5.

All the 'life-expectancies' quoted above are based on accumulating the current annual risks of dying – the force of mortality at the moment – and do not take into account any future progress. If we're going to say something about the prospects for children born now, or yet to be born, we have to make some assumptions about what is going to happen to human health and lifespan. The 'principal projections' for England and Wales estimate that males born now, allowing for projected future improvements in health, can *on average* expect to live until 90, and for females it's 94:[7] 32 per cent of males and 39 per cent of females are expected to reach 100 and get a letter from the Queen, or whoever has the job in the early twenty-second century.[8] Babies born in 2050 are projected to live until an average of 97 if male and 99 if female, meaning they will die around 2150. Each generation has an ever-longer reach into the future.

But these projections are, rather understandably, deeply controversial. Is there some inbuilt ageing process, and can it be reversed? Or is there some natural ceiling that we are going to bang our wrinkled, bald heads against? The arguments are fierce. Put a group of specialists on ageing together in a room, and you will be lucky if there are any survivors. People have claimed for years that a ceiling is being reached, and yet life-expectancy just keeps going steadily up. Individual extremes catch attention: Jeanne Calment was born in 1875, soon after the Franco-Prussian War, but kept soldiering on until 1997, when she died at 122, and there are now more verified 115-year-olds than you can shake a walking-stick at.

One thing is certain for children born in the future – there are going to be a lot of old people around for them to look after, as those already plodding through their middle age are not going to go away. The UN estimates that the proportion of people over 60 will double between 2007 and 2050, as people will live longer and lower fertility rates mean fewer young people.[9] There will be 2 billion people over 60 in the world by 2050, and around 400 million people over 80.

But what sort of state are all these old people going to be in? After the bit about three-score years and ten, Psalm 90 continues with 'and if by reason of strength they be fourscore years, yet is their strength labour and sorrow; for it is soon cut off, and we fly away', which doesn't exactly paint an encouraging picture of ageing. Are we making all this effort to live longer just so that we can spend even more time sitting round the edge of a room, television blaring, struggling to grasp what reluctant visitors are shouting at us?

Recent studies have suggested a more positive view of health: 78 per cent of an unselected group of over a thousand 85-year-olds in Newcastle upon Tyne reported their health as at least 'good' compared with others of their age,[10] and only 16 per cent suffered from depression.

But 61 per cent lived alone, and loneliness is associated with failing mental abilities, probably feared even more than physical disability. This is already a big issue and won't become any easier: 30 per cent of over-90s have dementia, and the number of people in Britain who are over 90 is predicted to rise from 1.2 million in 2010 to 1.8 million by 2051.[11] In 2005 it was estimated that 700,000 people were living with dementia in the UK, and this is likely to rise to 1.7 million by 2051, nearly all of whom are alive now.[12] Perhaps you are one of them.

In 1958 Paul McCartney saw 64 as near the end of life, whereas baby-boomers who were born just after the Second World War and have now reached this age tend to have a reasonable lifestyle and only see themselves as 'middle-aged', with much more before them than a precipitate decline. In 2008 UK men at age 65 had around 17 years of life expectancy left, 10 of which would be considered 'healthy', while women had more than 20 years left on average, 11 of which would be 'healthy'[13] – 'healthy' means that people consider their health as 'good' or 'very good' on a 5-point scale, which is quite a stringent criterion.

So what should these people be getting up to? Oddly, the proximity of death makes risk less risky. That sounds strange. But if you looked simply at relative risk probabilities, you could almost conclude that the old should be out there playing with fire. Here's how that surprising calculation works.

Recently DS was asked to do a tandem sky-dive for a TV programme and, of course, checked the data. Sky-diving is on average around 10 MicroMorts (see Figure 36), but he reckoned tandem would be a bit safer (only 1 death in 340,000 tandem jumps recorded by the British Parachute Association[14]). So he reckoned around 7 MicroMorts for the plane ride and jump.

For a young lad of 18 (average total annual risk of death from all causes 530 MicroMorts) this is equal to around 5 days of average background risk, whereas for a man DS's age (all-cause annual risk 7,000 MicroMorts) it is only around 9 hours' worth. So, from a relative perspective, it makes more sense for an old codger to hurl himself out of a plane, charge around on a Harley-Davidson and play chicken on motorways than some cocky youth with their life ahead of them. But try telling that to the youth.

JUDGEMENT DAY

W<small>E – THAT IS, THE AUTHORS</small> – have a problem: we like numbers and we think they matter, but we like our characters too. Call this vain – we invented them, after all – but Norm, Prudence and the Kevlins are mostly all right, we think. More than that, although we recognise the patterns in numbers as well as anyone, we can't bring ourselves to say that any of our characters is irrational by choosing to ignore them and going their own way.

It's become fashionable to blame people for their cognitive biases or failures of reasoning, but we think that much of what it's tempting to call irrationality about danger is really a result of the way the information is framed, or the sheer complexity of the decision, and that their reasoning is usually reasonable, for all their human ways.

It's not reasoning that we would necessarily share, but nor is it reasoning we can easily dismiss. You can call Norm a berk if you like, Kelvin a smug, offensive slug and Prudence a pain in the arse – though we think they deserve more understanding than that – but given what they want from life, we don't know how to prove that any of them is actually *wrong* about the choices they make in the face of danger. We don't know how to use data to tell them how to live. Even if planes feel more dangerous to travel in than cars, it's not much use to point out to any individual that the probabilities of fatality are the other way round, since what they mean by danger is seldom as simple as a mortality rate. It is often, as we've said, not even about danger.

They may be odd, but they are not stupid, they know what they care about. They live in an uncertain world, where risks can change and no one knows which side of the odds they'll fall. Their hopes and fears are not like the unfounded fears of monsters under the bed; they are fears in a real and usually messy human context.

This is not to turn all aspects of danger into relativism. Our view is that planes really are safer than cars, on average. We simply argue that a measure of what people call risk must be a matter of personal value and personal framing. The objective numbers can't be separated from subjective perception. A risk stated as 1 in 400 can't help but focus attention on the 1. To say that the risk is the same if we describe it as a chance of 399 in 400 of nothing happening might be mathematically correct, but when people react to these different frames differently, that does not, in our view, demonstrate their irrationality. It demonstrates the importance both of the numbers and of perspective. Think of it like the view of the countryside from the city, compared with the view of the city from the countryside. Both country and city exist in the same proportions wherever you stand, but that doesn't mean that where you stand is irrelevant. The view matters. So do the proportions. Risk does not exist independently of the way that people see it. Nor can the way they see it float free of the numbers.

We could go further. We looked at a lot of numbers to write this book – and we mean *a lot* – and we have a geeky streak ourselves. Possibly you noticed. Maybe that leads you to expect that our private attitude to risk will be on reason's side. And on the whole it is. But while we think the numbers are vital, we are not piano keys either, and we share a deep sense of the uncertainties around data, statistics and evidence. Whether you think that chance really exists, or that your fate was cast in the Big Bang or that it's all fixed by a deity or three sisters spinning, measuring and cutting your fate, we all still have to deal with not knowing quite what will happen. We think there's more uncertainty than you'd think from the way people throw numbers around.

We think this especially because when you try to grab hold of probability it somehow slips through your fingers. It's hard to say what it really is for an individual. It's hard, too, to say how the average affects any individual.

Of course, some things are more likely than others: the chance of getting wet if it rains compared with being – alone of 7 billion people on the planet – the one to cop an asteroid. Ignore these differences and we would worry about you. It shouldn't have taken until the 1650s, when Blaise Pascal and Pierre de Fermat began a correspondence about dice, for people to get round to putting numerical probabilities on events. But although in some ways it is a simple idea, it has caused headaches ever since.

As an example, on 14 April 2012, DS saw a tip in a newspaper and put £2 on Cappa Bleu to win the Grand National at Aintree. The odds were 16 to 1, which meant a profit of £32 for a win, but could also be interpreted as meaning that the bookmaker thought there was a 'chance' of around 6 per cent that Cappa Bleu would win.

Previously, DS had been to see his general practitioner, who measured his blood pressure and cholesterol, tapped at his computer and duly announced a 12 per cent chance of a heart attack or stroke in the next ten years. Disturbing ... until the GP also said that this was less than average for a man of the Professor's age, who then irrationally cheered up at this change of framing. After a strong finish, Cappa Bleu came in fourth.

But what do all these probabilities mean? Philosophers and statisticians have argued for centuries and are far from agreement. Into this sometimes savage controversy what can we do in a short space but charge, prejudices blazing? Here are ours, which mirror how we feel about Norm, Kelvin and Prudence.

Traditionally, probabilities have been based on known physical properties and pure reasoning – for example, a coin has two sides, so tossing it has a 1-in-2 chance of coming down heads; throwing a dice has a 1-in-6 chance of a six; dealing one card from a (properly) shuffled standard deck of cards has a 1-in-52 chance of turning up the ace of spades. But this only works if we start with some idea of 'equally likely events', in the sense that all cards are somehow equally likely to turn up. But that requires us to say what 'likely' means, so we're back where we started. (And then, of course, in real life someone might cheat.)

Another idea of probability is to say that it is how often something happens when a similar situation is repeated a vast number of times, such as the proportion of people who reach 100 years of age. But apart from very special set-ups such as flipping coins, these exactly similar situations just don't occur: there is only one you to reach 100, or not, only one Grand National that DS could bet on in 2012, and only one DS to have a heart attack, or not. The idea that we all conform to the frequencies of the dead takes most of the life out of life. What was is not necessarily a good guide to what might be, let alone what will be.

As a way of dealing with the one-offs that we all like to think we are (yes, we are all individuals) some philosophers offer the idea of an intrinsic tendency for an event to occur, so that all the vastly complex aspects of DS's current and future existence combine to give some sort of 'propensity' for him to have a heart attack or stroke in the next ten years, and the doctor's '12 per cent chance' is an estimate of this. The idea of Cappa Bleu having some true underlying propensity to win is attractive but does not seem useful or provable.

We reject all these explanations and take a very pragmatic stance – that this '12 per cent chance of a heart attack' is not really DS's risk, and not even an estimate of some propensity of it. It is based on a few items of limited information, and should be treated like the 'probability' of Cappa Bleu winning – reasonable betting odds given current information. No more and no less.

Treating probability like a bet seems a cop-out, but it has powerful implications. It means that any number we claim for a 'probability' is constructed by us based on what we know. It is necessarily a judgement and does not exist as a property of the outside world. Risk, in this sense, is a measure of what we don't and can't know as much as a measure of what we can.

All of which forms part of a rather startling conclusion: that independent, objective probability, as Norm says at the last, doesn't exist.*

---

* We concede that at the sub-atomic level there may well be irreducible, unavoidable and inevitable chance, which means that 'determined probabilities', to use Stephen Hawking's expression, could be said to exist. But this does not seem particularly relevant to our judgement about who will win the Grand National.

Nor, as we say, does the average person exist to whom the average risk is supposed to apply. The average is an abstraction. The reality is variation. Poor Norm, he was a man in search of data, but perhaps the data were really in search of him. Was he ever really there? He had to disappear once he finally stopped believing in the norm.

In some ways probability is a bit like the certainty that there will be people, the story with which we began, but a certainty that tells you nothing about who. In other words, it tells you only a tiny part about you. The probability that there would be at least one person called Norm – high – is only a tiny part of the story about how this particular – infinitely improbable – Norm came into being.

So in practical terms, for the events of life in general, when we say a certain activity is dangerous and quote its risk as so many MicroMorts, these numbers should be considered only as reasonable betting odds given what we know. As soon as we know more – maybe the age of the person about to try base-jumping and whether they are sober, or how many pike there are in the reservoir, if they're near by, how fast they swim, when they last ate, whether the particular one that hasn't eaten lately still feels hungry and is near enough and fast enough and is the sort to recognise the edible potential of human flesh, even disguised by sagging underpants, *and* is a hard-nosed, vicious-enough, brave-enough sod to go for a lad's tackle – the risk changes, suggesting that the potential degree of refinement is often infinite. And this can just as easily be applied to things that have happened but we don't know about yet, such as giving the odds that Jack the Ripper was really the duke of Clarence. Or Queen Victoria, for that matter.

Nor can any single risk capture the full complexity of our feelings and judgements about nature or the economy – or let's go the whole hyperbolic hog and add the meaning of life – and so none of the natural, economic, lifestyle or other risks that we talk about can be explained except in the context of a vast swathe of other values. Our psychological reactions can be both optimistic (Kelvin and sex) and pessimistic (Norm and pike), and who knows in any instance which will apply? Similarly, no calculation of probability *times* consequence can tell you what weight you should attach to the consequences – in the unlikely event that you know them all – a weight that can only be for you to decide. And if half the calculation of a risk is infinitely variable, what is the objective answer to that calculation?

Probability sounds sensible enough, but whenever you reach for a firm and meaningful definition the concept loses shape – although it is a number, show us the scales or stick that you can measure it with. Egyptians, Greeks, Babylonians and others did amazing things with algebra, geometry, number theory and much more, but they never even got started on probability, and the omission is telling. DS says he has spent many years trying to work out why people find probability intuitively difficult and confusing. He has concluded that it is because probability is intuitively difficult and confusing. MB adds that he has often reported people's communication of risk and found the communicators don't really know what they are communicating. Just when anxious people most want clarity, they find muddle. There is a reason for that. It *is* a muddle. Norm, sadly, could never work that out.

The view that probability doesn't exist is unusual but moderately respectable.[1] It is also liberating, as it means that we are free to use a variety of metaphors and analogies when talking about risk, chance or probability, free to look at it from multiple perspectives and accept that perspective matters.

For example, a 12 per cent risk of a heart attack is often communicated as 'out of 100 men like you, in 10 years we expect 12 to have a heart attack or stroke'. But there are not 100 men like you, and the probability is not yours. A more gripping metaphor might be to say 'of 100 ways that things may turn out for you over the next 10 years, in 12 of them you will have a heart attack or stroke.*

---

* The Bank of England uses a similar metaphor when communicating uncertainty about its forecasts with fan charts, in which the line describing predicted economic growth or inflation blurs out into a huge fan of graded possibilities, expressing what might be expected 'if economic circumstances identical to today's were to prevail on 100 occasions'. You can ask a computer to play out these futures multiple times, a technique known as a 'Monte Carlo simulation', which started with the US project to build a hydrogen bomb. Similarly, there are 'ensembles' for weather forecasting in which a number of different predictions are based on

So, which of the 100 are you this time? And when Prudence wonders if she is the one in 100 and says 'What if ...?', has she a point? Or when Norm says she exaggerates, doesn't he have one too? And when Kelvin says 'so what?', is he wrong?

Perhaps you have made your peace with the dual nature of risk, chance and probability, and you are comfortable with its elusiveness from a personal point of view. But if Norm, Prudence and Kelvin are not wrong to say what they say – each different, each to their own view of risk – then which is the right number for them? On the other hand, if you do happen to look up and see a falling piano ...

---

slightly perturbed assumptions about what is happening now, and chaos means that these small differences may result in wildly different predictions after a few days. Unfortunately there is still a reluctance to talk in public about the chances of different weather patterns, although in the US 'possible paths' of hurricanes are routinely shown on public news broadcasts.

Figure 36: **The authors' calculations of MicroMorts per hazard**

Cause of death	Context	Micromorts	Exposure	Source
'Non-natural causes'	E+W, 2010	1	Per day	(a)
Accident at birth	E+W, 2010	279	Per birth	(b)
Infant mortality (first year)	E+W, 2010	4,300	Per birth	(b)
Infant mortality	World, 2010	40,000	Per birth	(c)
Infant mortality	Sierra Leone, 2010	119,000	Per birth	(c)
Accidents – under 14	E+W, 2010	18	Per year	(a)
Accidental strangulation – under 14		3	Per year	(a)
Pedestrians – under 14	E+W, 2010	2	Per year	(a)
Murder/homicide	E+W, 2010	14	Per year	(d)
Murder/homicide – under 1	E+W, 2010	27	Per year	(d)
Murder/homicide – age 10–14	E+W, 2010	2	Per year	(d)
Murder/homicide – for black male	E+W, 2010	76	Per year	(d)
Measles	UK, 1960	200	Per illness	(e)
Anaesthesia for non-emergency operation	UK	10	Per operation	(f)
Giving birth	UK, 2010	120	Per birth	(g)
Giving birth	World, 2010	2,100	Per birth	(g)
Caesarean section	UK	170	Per birth	(f)
Death in hospital due to avoidable safety lapse	UK, 2010	76 (minimum)	Per day in hospital	(h)
Coronary artery bypass graft	UK, 2008	16,000	Per operation	(j)
Serving in Afghanistan (peak risk period)	UK forces, May–Oct 2009	47	Per day	(k)
Flying in Bomber Command in Second World War	RAF, 1939–45	25,000	Per mission	(l)
Walking	UK, 2010	1	Per 27 miles	(m)
Cycling	UK, 2010	1	Per 28 miles	(m)
Riding a motorbike	UK, 2010	1	Per 7 miles	(m)
Driving	UK, 2010	1	Per 333 miles	(m)
Train	UK, 2010	1	Per 7,500 miles	(n)

Cause of death	Context	Micromorts	Exposure	Source
Commercial aircraft	US, 1992–2011	1	Per 7,500 miles	(p)
Light aircraft ('general aviation')	US, 1992–2011	1	Per 15 miles	(p)
Scuba-diving	UK, 1998–2009	5	Per dive	(q)
Hang-gliding	UK	8	Per jump	(r)
Rock climbing	UK	3	Per climb	(r)
Sky-diving	UK	10	Per jump	(s)
Running marathon	US, 1975–2004	7	Per run	(t)
Ecstasy/MDMA (mentioned on death certificate)	E+W, 2003–7	1.7	Per week	(u)
Heroin (mentioned on death certificate)	E+W, 2003–7	377 (underestimate)	Per week	(u)
Asteroid	World	1	Per lifetime	(v)
Coalmining	UK, 1911	1,190	Per year	
Commercial fishing	UK, 1996–2005	1,020	Per year	
Coalmining	UK, 2006–10	430	Per year	
All occupations	UK, 2010	6	Per year	
All occupations	World	160	Per year	

Sources: (a) Office for National Statistics. Mortality Statistics: Deaths registered in England and Wales (Series DR). 2011 Available from: http://www.ons.gov.uk/ons/rel/vsob1/mortality-statistics--deaths-registered-in-england-and-wales--series-dr-/2010/index.html [accessed 28 Nov 2011]. (b) Office for National Statistics. Child mortality statistics: Childhood, infant and perinatal. 2012. Available from: http://www.ons.gov.uk/ons/rel/vsob1/child-mortality-statistics--childhood--infant-and-perinatal/2010/index.html [accessed 9 May 2012]. (c) UN Inter-Agency Group for Child Mortality Estimation. Child Mortality Estimates. 2012. Available from: http://www.childmortality.org/ [accessed 9 May 2012]. (d) Home Office. Homicides, Firearm Offences and Intimate Violence 2010/11: Supplementary Vol. 2 to Crime in England and Wales. 2012. Available from: http://www.homeoffice.gov.uk/publications/science-research-statistics/research-statistics/crime-research/hosb0212/ [accessed 4 Nov 2012]. (e) Health Protection Agency. Measles notifications and deaths in England and Wales, 1940–2008. 2012. Available from: http://www.hpa.org.uk/web/HPAweb&HPAwebStandard/HPAweb_C/1195733835814 [accessed 13 May 2012]. (f) Royal College of Anaesthetists. Death or brain-damage. 2009. Available from: http://www.rcoa.ac.uk/document-store/death-or-brain-damage [accessed 30 Oct 2012]. (g) World Health Organisation. Trends in maternal mortality: 1990–2010. WHO. 2012. Available from: http://www.who.int/reproductivehealth/publications/monitoring/9789241503631/en/index.html [accessed 14 Nov 2012]. (h) National Patient Safety Agency. Quarterly Data Summaries. Available from: http://www.nrls.npsa.nhs.uk/resources/collections/quarterly-data-summaries/?entryid45=133687 [accessed 29 Jan 2013]. (j) Care Quality Commission. Survival Rates – Heart Surgery in United Kingdom, 2008–9. 2009. Available from: http://heartsurgery.cqc.org.uk/survival.aspx [accessed 25 Nov 2012]. (k) Bird, S., Fairweather, C. Recent military fatalities in Afghanistan by cause and nationality: Period 15, 5 September 2011 to 22 January 2012. Available from: http://www.mrc-bsu.cam.ac.uk/Publications/PDFs/PERIOD_15_fatalities_in_Afghanistan_by_cause_and_nationality.pdf [accessed 16 Nov 2012]. (l) Wikipedia. RAF Bomber Command. Wikipedia, the free encyclopedia. 2013. Available from: http://en.wikipedia.org/w/index.php?title=RAF_Bomber_Command&oldid=531643454 [accessed 29 Jan 2013]. (m) Department for Transport. Reported road casualties Great Britain: main results 2010. 2011. Available from: http://www.dft.gov.uk/statistics/releases/reported-road-casualties-gb-main-results-2010 [accessed 28 Nov 2011]. (n) Rail Safety and Standards Board. Annual Safety Performance Report 2010/11. 2011. Available from: http://www.rssb.co.uk/SPR/REPORTS/Documents/ASPR%202010-11%20Final.pdf. (p) NTSB National Transportation Safety Board. NTSB – Aviation Statistical Reports. Available from: http://www.ntsb.gov/data/aviation_stats.html [accessed 29 Oct 2012]. (q) British Sub-Aqua Club. UK Diving Fatalities Review. Available from: http://www.bsac.com/page.asp?section=3780&sectionTitle=UK+Diving+Fatalities+Review [accessed 20 Jan 2012]. (r) Health Safety Executive. Risk education - Statistics. 2003. Available from: http://www.hse.gov.uk/education/statistics.htm [accessed 26 Nov 2012]. (s) British Parachute Association. How Safe. 2012. Available from: http://www.bpa.org.uk/staysafe/how-safe/ [accessed 26 Nov 2012]. (t) Redelmeier, D.A., Greenwald, J.A. Competing Risks of Mortality with marathons: retrospective analysis. BMJ. 2007 Dec 22;335(7633):1275–7. (u) King, L.A., Corkery, J.M., An Index of Fatal Toxicity for Drugs of Misuse. Hum Psychopharmacol. 2010 Mar;25(2):162–6; Home Office. Drug Misuse Declared: Findings from the 2010/11 British Crime Survey England and Wales. 2011. Available from: http://www.homeoffice.gov.uk/publications/science-research-statistics/research-statistics/crime-research/hosb1211/ [accessed 14 Nov 2012]. (v) National Research Council. Defending Planet Earth: Near-Earth Object Surveys and Hazard Mitigation Strategies (Washington, D.C.: The National Academies Press, 2010).

Figure 37: **Average MicroLives (1/2 hour of life-expectancy) lost or gained per day of exposure to a specified hazard ratio for all-cause mortality, averaged over life after 35**

Factor	Definition of daily exposure	Males over 35 Estimated change in life-expectancy (years)	Males over 35 Micro-Lives per day	Females over 35 Estimated change in life-expectancy (years)	Females over 35 Micro-Lives per day
Smoking	Smoking 15–24 cigarettes (a)	– 7.7	– 10	– 7.3	– 9
Alcohol	First drink (10 g alcohol) (b)	1.1	1	0.9	1
	Each subsequent drink (up to 6)	–0.7	– ½	–0.6	– ½
Obesity	BMI: per 5 kg above 22.5 (c)	– 2.5	– 3	– 2.4	– 3
	per 5kg above optimum weight for average height	– 0.8	– 1	– 0.9	– 1
Sedentary behaviour	2 hours watching television (d)	–0.7	– 1	– 0.8	– 1
Red meat	One portion (85 g, 3 oz) (e)	–1.2	– 1	– 1.2	– 1
Fruit and vegetable intake	Five servings or more (blood vitamin C > 50 nmol/l) (f)	4.3	4	3.8	4
Coffee	2–3 cups (g)	1.1	1	0.9	1
Physical activity	First 20 minutes of moderate exercise (h)	2.2	2	1.9	2
	Subsequent 40 minutes of moderate exercise	0.7	1	0.5	½
Statins	Taking a statin (j)	1	1	0.8	1
Air pollution	Living in Mexico City rather than London (k)	0.6	– ½	0.6	–½
Gender	Being male rather than female (l)	– 3.7	– 4	–	–
Geography	Resident of Sweden vs Russia (m)	– 14.1	– 21	– 7.6	– 9
Era	Living in 2010 vs. 1910 (m)	13.5	15	15.2	15
	Living in 2010 vs. 1980	7.5	8	5.2	5
Single dose of ionising radiation	0.07 mSv (e.g., single transatlantic flight)	30 mins	–1	30 mins	–1

Sources: (a) Doll, R. Mortality in Relation to Smoking: 50 Years' Observations on Male British Doctors. *BMJ*. 2004 Jun 26;328:1519–20. (b) Di Castelnuovo, A., Costanzo, S., Bagnardi, V., Donati, M.B., Iacoviello, L., De Gaetano, G. Alcohol Dosing and Total Mortality in Men and Women: An Updated Meta-Analysis of 34 Prospective Studies. *Arch. Intern. Med.* 2006 Dec 11;166(22):2437–45. (c) Prospective Studies Collaboration. Body-Mass Index and Cause-Specific Mortality in 900,000 Adults: Collaborative Analyses of 57 Prospective Studies. *The Lancet*. 2009 Mar;373:1083–96. (d) Wijndaele, K., Brage, S., Besson, H., Khaw, K.-T., Sharp, S.J., Luben, R., et al. Television Viewing Time Independently Predicts All-Cause and Cardiovascular Mortality: The EPIC Norfolk Study. *Int. J. Epidemiol.* 2011 Feb 1;40(1):150–9. (e) Pan, A., Sun, Q., Bernstein, A.M., Schulze, M.B., Manson, J.E., Stampfer, M.J., et al. Red Meat Consumption and Mortality: Results from 2 Prospective Cohort Studies. *Arch Intern Med*. 2012 Apr 9;172(7):555–63. (f) Khaw, K.-T., Wareham, N., Bingham, S., Welch, A., Luben, R., Day, N. Combined Impact of Health Behaviours and Mortality in Men and Women: The EPIC-Norfolk Prospective Population Study. *PLoS Med*. 2008 Jan 8;5(1):e12. (g) Freedman, N.D., Park, Y., Abnet, C.C., Hollenbeck, A.R., Sinha, R. Association of Coffee Drinking with Total and Cause-Specific Mortality. *N. Engl. J. Med.* 2012 May 17;366(20):1891–904. (h) Woodcock, J., Franco, O.H., Orsini, N., Roberts, I. Non-Vigorous Physical Activity and All-Cause Mortality: Systematic Review and Meta-Analysis of Cohort Studies. *Int J Epidemiol.* 2011 Feb;40(1):121–38. (j) Ray, K.K., Seshasai, S.R.K., Erqou, S., Sever, P., Jukema, J.W., Ford, I., et al. Statins and All-Cause Mortality in High-Risk Primary Prevention:

A Meta-Analysis of 11 Randomized Controlled Trials Involving 65,229 Participants. *Arch. Intern. Med.* 2010 Jun 28;170(12):1024–31. (*k*) Pope, C.A., Ezzati, M., Dockery, D.W. Fine-Particulate Air Pollution and Life Expectancy in the United States. *New England Journal of Medicine.* 2009;360(4):376–86. (*l*) Office for National Statistics. Interim Life Tables, 2008–10. Available from: http://www.ons.gov.uk/ons/rel/lifetables/interim-life-tables/2008-2010/index.html [accessed 13 Feb 2012]. (*m*) Human Mortality Database. 2012. Available from: http://www.mortality.org/ [accessed 8 Aug 2012].

NOTES

Note: All web addresses were live when last accessed 8 April, 2013.

Introduction

1. Gilbert, M., Busund, R., Skagseth, A., Nilsen, P.Å., Solbø, J.P. 'Resuscitation from accidental hypothermia of 13·7°C with circulatory arrest'. *The Lancet.* 2000 Jan;355(9201):375–6.
2. Gawande A. *Better: A Surgeon's Notes on Performance* (London: Profile Books, 2008).
3. Office for National Statistics. Mortality Statistics: Deaths Registered in England and Wales (Series DR). Available from: http://www.ons.gov.uk/ons/rel/vsob1/mortality-statistics--deaths-registered-in-england-and-wales--series-dr-/2010/index.html.

Chapter 1: The beginning

1. NASA. Asteroid 2008 TC3 Strikes Earth: Predictions and Observations Agree. Available from: http://neo.jpl.nasa.gov/news/2008tc3.html.
2. NHS Maternity Statistics – England, 2011–2012. Table 30: Median Birth Weight (Grams) of Live Born Singleton and Multiple Deliveries. http://www.hscic.gov.uk/catalogue/PUB09202.
3. Howard, R.A. 'Microrisks for Medical Decision Analysis'. *International Journal of Technology Assessment in Health Care.* 1989;5(03):357–70.
4. Office for National Statistics. Mortality Statistics: Deaths Registered in England and Wales (Series DR). Available from: http://www.ons.gov.uk/ons/rel/vsob1/

mortality-statistics--deaths-registered-in-england-and-wales--series-dr-/2010/
index.html.

5. Royal College of Anaesthetists. Death or Brain Damage. 2009. Available
from: http://www.rcoa.ac.uk/document-store/death-or-brain-damage.

6. Bird, S., Fairweather, C. Recent Military Fatalities in Afghanistan (and Iraq)
by Cause and Nationality. 2010. Available from: http://www.mrc-bsu.cam.
ac.uk/Publications/PDFs/PERIOD 9_10_fatalities_in_Afghanistan_and_
Iraq.pdf.

7. Department for Transport GMH. Transport Analysis Guidance – WebTAG.
2009. Available from: http://www.dft.gov.uk/webtag/documents/expert/
unit3.4.1.php.

Chapter 2: Infancy

1. Didion, J. *Blue Nights* (New York: Knopf Doubleday, 2011).

2. Office for National Statistics. UK Interim Life Tables, 1980–82 to 2008–10.
2011. Available from: http://www.ons.gov.uk/ons/rel/lifetables/interim-life-
tables/2008-2010/index.html.

3. Schellekens, J. 'Economic Change and Infant Mortality in England,
1580–1837'. *Journal of Interdisciplinary History*. 2001;32(1):1–13.

4. Office for National Statistics. Child Mortality Statistics: Childhood, Infant
and Perinatal. 2012. Available from: http://www.ons.gov.uk/ons/rel/vsob1/
child-mortality-statistics--childhood--infant and-perinatal/2010/index.html.

5. Ibid.

6. National Perinatal Mortality Unit. The Birthplace Cohort Study: Key
Findings. 2012. Available from: https://www.npeu.ox.ac.uk/birthplace/
results.

7. Office for National Statistics. Unexplained Deaths in Infancy: England and
Wales, 2009. Available from: http://www.ons.gov.uk/ons/rel/child-health/
unexplained-deaths-in-infancy--england-and-wales/2009/new-component.
html.

8. UN Inter-Agency Group for Child Mortality Estimation. Child Mortality
Estimates. 2012. Available from: http://www.childmortality.org/.

9. UNICEF. Childinfo.org: Statistics by Area – Child Mortality – Infant
Mortality [Internet]. 2012. Available from: http://www.childinfo.org/
mortality_imrcountrydata.php.

10. See note 4 above.

11. United Nations Development Programme. MDG MONITOR :: Goal ::
Reduce Child Mortality. Available from: http://www.mdgmonitor.org/
goal4.cfm.

12. Rudski, J.M., Osei, W., Jacobson, A.R., Lynch, C.R. 'Would You Rather Be Injured by Lightning or a Downed Power Line? Preference for Natural Hazards.' *Judgment and Decision Making*. 2011;6(4):314–22.

Chapter 3: Violence

1. Best, J. *Threatened Children: Rhetoric and Concern about Child-Victims* (Chicago: University of Chicago Press, 1993).
2. Home Office. Homicides, Firearm Offences and Intimate Violence 2010/11: Supplementary Volume 2 to Crime in England and Wales. 2012. Available from: https://www.gov.uk/government/publications/homicides-firearm-offences-and-intimate-violence-2010-to-2011-supplementary-volume-2-to-crime-in-england-and-wales-2010-to-2011.
3. BBC News. Child killing rate 'drops by 40%' in England and Wales. 2010. Available from: http://news.bbc.co.uk/1/hi/uk/8497277.stm.
4. Rooney, C., Devis, T. Recent Trends in Deaths from Homicide in England and Wales: Health Statistics Quarterly Autumn 2009. 1999. Available from: http://www.ons.gov.uk/ons/rel/hsq/health-statistics-quarterly/no--3--autumn-1999/index.html; NSPCC; Child Killings in England and Wales: Explaining the Statistics. 2012. Available from: http://www.nspcc.org.uk/Inform/research/briefings/child_killings_in_england_and_wales_wda67213.html#Homicide_statistics.
5. Newiss, G. Child Abduction: Understanding Police Recorded Crime Statistics. 2008. Available from: http://www.chimat.org.uk/resource/item.aspx?RID=62767.
6. CEOP. Scoping Report on Missing and Abducted Children. 2011. Available from: http://ceop.police.uk/Documents/ceopdocs/Missing_scopingreport_2011.pdf.
7. Ministry of Justice. MAPPA Reports. 2012. Available from: http://www.justice.gov.uk/statistics/mappa-reports.
8. Cohen, S. *Folk Devils and Moral Panics*, 30th anniversary edn (London: Routledge, 2002).

Chapter 4: Nothing

1. *Daily Express.* Daily Fry-Up Boosts Cancer Risk by 20 Per Cent. Available from: http://www.express.co.uk/posts/view/295296/Daily-fry-up-boosts-cancer-risk-by-20-per-cent.
2. *Daily Telegraph.* Nine in 10 People Carry Gene Which Increases Chance of High Blood Pressure. Available from: http://www.telegraph.co.uk/health/healthnews/4630664/Nine-in-10-people-carry-gene-which-increases-chance-of-high-blood-pressure.html.

3. Woolf, V. *The Common Reader*, 1st ser., annotated edn (Orlando, FL: Houghton Mifflin Harcourt, 2002).

4. Kahneman, D. *Thinking, Fast and Slow* (New York: Farrar, Straus and Giroux, 2011).

5. Harrabin, R., Coote, A., Allen, J. *Health in the News: Risk, Reporting and Media Influence* (London: King's Fund, 2003).

Chapter 5: Accidents

1. Office for National Statistics. Mortality Statistics: Deaths Registered in England and Wales (Series DR). 2011. Available from: http://www.ons.gov.uk/ons/rel/vsob1/mortality-statistics--deaths-registered-in-england-and-wales--series-dr-/2010/index.html.

2. *Daily Telegraph*. Girl Cannot Walk To Bus Stop Alone. 2010. Available from: http://www.telegraph.co.uk/news/uknews/8001444/Girl-cannot-walk-to-bus-stop-alone.html.

3. Department for Transport. Reported Road Casualties Great Britain: Main Results 2010. 2011. Available from: http://www.dft.gov.uk/statistics/releases/reported-road-casualties-gb-main-results-2010.

4. Savage, L.J. 'The Theory of Statistical Decision'. *Journal of the American Statistical Association*. 1951 Mar 1;46(253):55–67.

5. Office for National Statistics. Avoidable Mortality in England and Wales. 2012. Available from: http://www.ons.gov.uk/ons/rel/subnational-health4/avoidable-mortality-in-england-and-wales/2010/index.html.

6. Central Statistical Office. *Annual Statistical Abstract 1951*.

7. Moran, J. 'Crossing the Road in Britain, 1931–1976'. *The Historical Journal*. 2006;49(02):477–96.

8. See note 3 above.

9. See note 1 above.

10. *The Guardian*. Ikea Recalls Over 3 Million Window Blinds, Shades. 2010. Available from: http://www.guardian.co.uk/world/feedarticle/9121407.

11. Gill, T. *No Fear* (London: Calouste Gulbenkian Foundation, 2007). Available from: http://www.gulbenkian.org.uk/publications/publications/42-NO-FEAR.html.

12. Health and Safety Executive. Children's Play and Leisure – Promoting a Balanced Response. 2012. Available from: http://www.hse.gov.uk/entertainment/childrens-play-july-2012.pdf.

13. Countryside Alliance. Outdoor Education – the Countryside as a Classroom. Available from: http://www.countryside-alliance.org/ca/campaigns-education/give-children-the-opportunity-to-learn-outside-of-the-classroom.

14. Health and Safety Executive. HSE – School Trips – Glenridding Beck – 10 Vital Questions. 2005. Available from: http://www.hse.gov.uk/services/education/school-trips.htm#statistics.
15. Rainey, S. 'Kellogg's Adds Vitamin D to Cereal to Fight Rickets'. *Daily Telegraph*. Available from: http://www.telegraph.co.uk/health/healthnews/8854634/Kelloggs-adds-vitamin-D-to-cereal-to-fight-rickets.html.

Chapter 6: Vaccination

1. Vaccine Liberation Army. Armed with Knowledge. 2012. Available from: http://vaccineliberationarmy.com/.
2. Department of Health. A Guide to Immunisations up to 13 Months of Age. 2007. Available from: http://www.nhs.uk/Planners/vaccinations/Documents/A%20guide%20to%20immunisations%20up%20to%2013%20months%20of%20age.pdf.
3. Carrington, Tammy. A Vaccination Horror Story. Available from: http://www.drlwilson.com/articles/VACCINE%20HORROR.htm.
4. BBC News. Measles Outbreak Prompts Plea To Vaccinate Children. 2011. Available from: http://www.bbc.co.uk/news/health-13561766.
5. Centers for Disease Control and Prevention. School and Childcare Vaccination Surveys. 2012. Available from: http://www2a.cdc.gov/nip/schoolsurv/schImmRqmt.asp.
6. Health Protection Agency. Measles Notifications and Deaths in England and Wales, 1940–2008. 2012. Available from: http://www.hpa.org.uk/web/HPAweb&HPAwebStandard/HPAweb_C/1195733835814.
7. Centers for Disease Control and Prevention. Pinkbook: Measles Chapter – Epidemiology of Vaccine-Preventable Diseases. 2012. Available from: http://www.cdc.gov/vaccines/pubs/pinkbook/meas.html#complications.
8. De Martel, C., Ferlay, J., Franceschi, S., Vignat, J., Bray, F., Forman, D., et al. 'Global Burden of Cancers Attributable to Infections in 2008: A Review and Synthetic Analysis.' *The Lancet* Oncology. May 2012. Available from: http://www.thelancet.com/journals/lanonc/article/PIIS1470-2045(12)70137-7/abstract.
9. NHS Information Centre. NHS Immunisation Statistics, England 2009–10. 2012. Available from: http://www.ic.nhs.uk/Article/1685.
10. Medicines and Healthcare Products Regulatory Agency (MHRA). Human Papillomavirus (HPV) Vaccine. 2012. Available from: http://www.mhra.gov.uk/PrintPreview/DefaultSplashPP/CON023340?ResultCount=10&DynamicListQuery=&DynamicListSortBy=xCreationDate&Dynamic

ListSortOrder=Desc&DynamicListTitle=&PageNumber=1&Title=Hu
man%20papillomavirus%20(HPV)%20vaccine.

11. Centers for Disease Control and Prevention. Vaccination Side Effects: HPV-
Cervarix. 2012. Available from: http://www.cdc.gov/vaccines/vac-gen/side-
effects.htm#hpvcervarix.

12. *Daily Mail*. Schoolgirl, 14, Dies After Being Given Cervical Cancer Jab.
2012. Available from: http://www.dailymail.co.uk/news/article-1216714/
Schoolgirl-14-dies-given-cervical-cancer-jab.html.

13. GPonline. Malignant Tumour Caused HPV Jab Girl's Death. 2012.
Available from: http://www.gponline.com/News/article/942531/
Malignant-tumour-caused-HPV-jab-girls-death/.

14. Goldman, A.S., Schmalstieg, E.J., Freeman, D.H. Jr, Goldman, D.A.,
Schmalstieg, F.C. Jr. What Was the Cause of Franklin Delano Roosevelt's
Paralytic Illness? *J. Med Biogr*. 2003 Nov;11(4):232–40.

15. Centers for Disease Control and Prevention. Seasonal Influenza (Flu) –
Questions and Answers – Guillain-Barré Syndrome (GBS). Available from:
http://www.cdc.gov/flu/protect/vaccine/guillainbarre.htm.

16. Sencer, D.J., Millar, J.D. Reflections on the 1976 Swine Flu Vaccination
Program. *Emerging Infectious Diseases*. 2006 Jan;12(1):23–8.

17. Andrews, N., Stowe, J., Al-Shahi Salman, R., Miller, E., Guillain–Barré
Syndrome and H1N1 (2009) Pandemic Influenza Vaccination Using an AS03
Adjuvanted Vaccine in the United Kingdom: Self-Controlled Case Series.
Vaccine. 2011 Oct 19;29(45):7878–82.

18. Centers for Disease Control and Prevention. Mercury and Thimerosal –
Vaccine Safety. 2012. Available from: http://www.cdc.gov/vaccinesafety/
Concerns/thimerosal/.

19. World Health Organisation. Measles. 2012. Available from: http://www.who.
int/mediacentre/factsheets/fs286/en/index.html.

Chapter 7: Coincidence

1. Understanding Uncertainty. Wasted Stamp. 2012. Available from: http://
understandinguncertainty.org/user-submitted-coincidences/wasted-stamp.

2. Lodge, D. *The Art of Fiction* (London: Random House, 2011).

3. Understanding Uncertainty. Cambridge Coincidences Collection. 2012.
Available from: http://understandinguncertainty.org/coincidences.

4. Diaconis, P., Mosteller, F. Methods for Studying Coincidences. *Journal of the
American Statistical Association*. 1989;84(408):853–61.

5. Biological daughter. 2012. Available from: http://understandinguncertainty.
org/user-submitted-coincidences/biological-daughter.

6. Born in the same bed. 2012. Available from: http:/understandinguncertainty. org/user-submitted-coincidences/born-same-bed.

7. Junk Shop Find. 2012. Available from: http://understandinguncertainty.org/ user-submitted-coincidences/junk-shop-find-0.

8. Army Coat Hanger. 2012. Available from: http://understandinguncertainty. org/user-submitted-coincidences/army-coat-hanger.

9. Koestler, A. *The Case of the Midwife Toad*, illustrated edn (London: Hutchinson, 1971).

10. See note 4 above.

11. *Daily Mail*. Couple Gives Birth to Three Children on the Same Day ... 14 Years Apart. Mail Online. 2008. Available from: http://www.dailymail. co.uk/news/article-518525/Couple-gives-birth-children-day--14-years-apart. html.

12. BBC. Three Children, Same Birthday. 2010; Available from: http://news. bbc.co.uk/1/hi/wales/8511586.stm.

13. *Daily Mail*. Happy Birthday to You: Couple Have 3 Children All Born on Same Date. Mail Online. 2010. Available from: http://www.dailymail.co.uk/ news/article-1320113/Happy-birthday-Couple-3-children-born-date.html.

14. Public phone box. 2012. Available from: http://understandinguncertainty. org/user-submitted-coincidences/public-phone-box.

15. See note 4 above.

Chapter 8: Sex

1. The National Archives. Public Information Films, 1979 to 2006, Film index, AIDs Monolith. 2012. Available from: http://www.nationalarchives. gov.uk/films/1979to2006/filmpage_aids.htm; National Archives. Public Information Films, 1964 to 1979, Film index, Peach And Hammer. 2012. Available from: http://www.nationalarchives.gov.uk/films/1964to1979/ filmpage_hammer.htm; National Archives. Public Information Films, 1979 to 2006, Film index, Crime Prevention – Hyenas. 2102. Available from: http:// www.nationalarchives.gov.uk/films/1979to2006/filmpage_crime.htm.

2. Colombo, B., Masarotto, G. Daily Fecundability. *Demographic Research* (6 Sep 2000);3. Available from: http://www.demographic-research.org/ Volumes/Vol3/5/default.htm.

3. Leridon, H. Can Assisted Reproduction Technology Compensate for the Natural Decline in Fertility with Age? A Model Assessment. *Human Reproduction*. 2004 Jul 1;19(7):1548–53.

4. NHS Clinical Knowledge Summaries. Effectiveness of Contraceptives. 2012. Available from: http://www.cks.nhs.uk/clinical_topics/by_alphabet/c.

5. Ibid.

6. Rosling, H. Children Per Woman Updated. 2012. Available from: http://www.gapminder.org/data-blog/children-per-woman-updated/.

7. Department for Education. Under-18 and under-16 Conception Statistics. Available from: http://www.education.gov.uk/childrenandyoungpeople/healthandwellbeing/teenagepregnancy/a0064898/under-18-and-under-16-conception-statistics.

8. UNICEF. A League Table of Teenage Births in Rich Nations. 2001. Available from: http://www.unicef-irc.org/publications/328.

9. Varghese, B., Maher, J.E., Peterman, T.A., Branson, B.M., Steketee, R.W. Reducing the Risk of Sexual HIV Transmission: Quantifying the Per-Act Risk for HIV on the Basis of Choice of Partner, Sex Act, and Condom Use. *Sex Transm Dis.* 2002 Jan;29(1):38–43.

10. Platt, R., Rice, P.A., McCormack, W.M. Risk of Acquiring Gonorrhea and Prevalence of Abnormal Adnexal Findings among Women Recently Exposed to Gonorrhea. *JAMA*. 1983 Dec 16;250(23):3205–9; Holmes, K.K., Johnson, D.W., Trostle, H.J. An Estimate of the Risk of Men Acquiring Gonorrhea by Sexual Contact with Infected Females. *Am. J. Epidemiol.* 1970 Feb 1;91(2):170–74.

11. Nawrot, T.S., Perez, L., Künzli, N., Munters, E., Nemery, B. Public Health Importance of Triggers of Myocardial Infarction: A Comparative Risk Assessment. *The Lancet.* 2011 Feb;377(9767):732–40.

12. Blanchard, R., Hucker, S.J. Age, Transvestism, Bondage, and Concurrent Paraphilic Activities in 117 Fatal Cases of Autoerotic Asphyxia. *BJP.* 1991 Sep 1;159(3):371–7.

13. HM Treasury e-CT. Optimism Bias. Available from: http://www.hm-treasury.gov.uk/green_book_guidance_optimism_bias.htm.

14. Sharot, T. *The Optimism Bias: A Tour of the Irrationally Positive Brain* (New York: Random House, 2011).

Chapter 9: Drugs

1. Nutt, D. *Drugs Without the Hot Air: Minimising the Harms of Legal and Illegal Drugs* (Cambridge: Uit Cambridge Ltd, 2012).

2. Parssinen, T.M. *Secret Passions, Secret Remedies: Narcotic Drugs in British Society, 1820–1930* (Manchester: Manchester University Press, 1983).

3. Conan Doyle, A.C. *The Sign of Four* (London: Spencer Blackett, 1890).

4. Thompson, H.S., Torrey, B., Simonson, K. *Conversations with Hunter S. Thompson* (Jackson, MI: University Press of Mississippi, 2008).

5. Kemp, C. *Painkiller Addict: From Wreckage to Redemption – My True Story.* (London, Hachette, 2012).

6. Fielding, L. Why I've Come to Consider Again the Potential Problems of Cannabis. *The Guardian*. 2012. Available from: http://www.guardian.co.uk/commentisfree/2012/aug/28/why-changed-mind-about-cannabis.

7. Home Office. Drug Misuse Declared: Findings from the 2010/11 British Crime Survey England and Wales. Home Office. 2011. Available from: https://www.gov.uk/government/publications/drug-misuse-declared-findings-from-the-2010-11-british-crime-survey-england-and-wales--12.

8. Office for National Statistics. Deaths Related to Drug Poisoning in England and Wales. 2011. Available from: http://www.ons.gov.uk/ons/rel/subnational-health3/deaths-related-to-drug-poisoning/2010/index.html.

9. King, L.A., Corkery, J.M. An Index of Fatal Toxicity for Drugs of Misuse. *Hum. Psychopharmacol.* 2010 Mar;25(2):162–6.

10. Ibid.

11. Advisory Council on the Misuse of Drugs. Cannabis: Classification and Public Health. 2008. Available from: https://www.gov.uk/government/publications/acmd-cannabis-classification-and-public-health-2008.

12. Nutt, D.J., King, L.A., Phillips, L.D. Drug Harms in the UK: A Multicriteria Decision Analysis. *The Lancet.* 2010 Nov;376(9752):1558–65.

13. Nutt, D.J. Equasy – An Overlooked Addiction with Implications for the Current Debate on Drug Harms. *J. Psychopharmacol.* (Oxford). 2009 Jan;23(1):3–5.

14. Journal of Psychopharmacology January 2009 23: 3–5. Also available from: http://www.encod.org/info/equasy-a-harmful-addiction.html.

Chapter 10: Big Risks

1. *The Guardian*. Sharp Decline in Public's Belief in Climate Threat, British Poll Reveals. 2010. Available from: http://www.guardian.co.uk/environment/2010/feb/23/british-public-belief-climate-poll.

2. Kahan, D.M., Jenkins-Smith, H., Braman, D. Cultural Cognition of Scientific Consensus. SSRN eLibrary. 7 Feb 2010. Available from: http://papers.ssrn.com/sol3/papers.cfm?abstract_id=1549444.

3. Kahan, D.M. et al. Affect, Values, and Nanotechnology Risk Perceptions: An Experimental Investigation. Yale Law School, Public Law Working Paper No. 155. Available from: http://papers.ssrn.com/sol3/papers.cfm?abstract_id=968652##.

4. Ibid.

5. Douglas, M., Wildavsky, A. *Risk and Culture: An Essay on the Selection of Technological and Environmental Dangers* (Berkeley and Los Angeles, University of California Press, 1983).

6. Haidt, J. *The Righteous Mind: Why Good People Are Divided by Politics and Religion* (New York: Pantheon Books, 2012).

7. Cabinet Office. National Risk Register. 2012. Available from: https://www.gov.uk/government/publications/national-risk-register-of-civil-emergencies.

8. Centers for Disease Control and Prevention. Crisis & Emergency Risk Communication. 2012 Available from: http://www.bt.cdc.gov/cerc/.

9. Wikipedia encyclopedia. 2011 Germany E. coli O104:H4 outbreak.. 2012. Available from: http://en.wikipedia.org/wiki/2011_Germany_E._coli_O104:H4_outbreak.

10. Hall, S.S. Scientists on Trial: At Fault? *Nature.* 2011 Sep 14;477(7364):264–9.

Chapter 11: Giving Birth

1. Rappaport, H. *Queen Victoria: A Biographical Companion* (Santa Barbara, CA: ABC-CLIO, 2003).

2. Byatt, A.S. *Still Life* (New York: Scribner's, 1997).

3. World Health Organisation. Trends in Maternal Mortality: 1990 to 2010. 2012. Available from: http://www.who.int/reproductivehealth/publications/monitoring/9789241503631/en/index.html.

4. Bird, S., Fairweather, C. Recent Military Fatalities in Afghanistan by Cause and Nationality· Period 15, 5 September 2011 to 22 January 2012. Available from: http://www.mrc-bsu.cam.ac.uk/Publications/PDFs/PERIOD_15_fatalities_in_Afghanistan_by_cause_and_nationality.pdf.

5. Wikipedia encyclopedia. Historical Mortality Rates of Puerperal Fever. 2012. Available from: http://en.wikipedia.org/w/index.php?title=Historical_mortality_rates_of_puerperal_fever&oldid=516214953.

6. Office for National Statistics. Mortality Statistics: Deaths Registered in England and Wales (Series DR). 2011. Available from: http://www.ons.gov.uk/ons/rel/vsob1/mortality-statistics--deaths-registered-in-england-and-wales--series-dr-/2010/index.html.

7. Centre for Maternal and Child Enquiries. Saving Mothers' Lives: Reviewing Maternal Deaths to Make Motherhood Safer: 2006–2008. *BJOG: An International Journal of Obstetrics & Gynaecology.* 2011;118:1–203.

8. Patient.co.uk. Maternal Mortality. 2012. Available from: http://www.patient.co.uk/doctor/Maternal-Mortality.htm.

9. See note 3 above.

10. Royal College of Anaesthetists. Death or Brain Damage. 2009. Available from: http://www.rcoa.ac.uk/document-store/death-or-brain-damage.

Chapter 12: Gambling

1. King James Bible (internet version). Available from: http://www.kingjamesbibleonline.org/.
2. Stone, P. *The Luck of the Draw:The Role of Lotteries in Decision Making* (Oxford: Oxford University Press, 2011).
3. Fienberg, S.E. Randomization and Social Affairs: The 1970 Draft Lottery. *Science.* 1971 Jan 22;171(3968):255–61.
4. David, F.N. *Games, Gods, and Gambling: A History of Probability and Statistical Ideas* (New York: Dover Publications, 1998).
5. New Living Translation. 2012. Available from: http://www.newlivingtranslation.com/.
6. British History Online. Elizabeth – July 1588, 1–5. Calendar of State Papers Foreign, Elizabeth, Vol. 22. 2003. Available from: http://www.british-history.ac.uk/report.aspx?compid=74849.
7. Hacking, I. *The Emergence of Probability*, 2nd edn (New York: Cambridge University Press, 2006).
8. Cardano, G. *Liber de ludo aleae* (Rome: FrancoAngeli, 2006).
9. See note 4 above.
10. Reith, G. *The Age of Chance: Gambling in Western Culture* (London: Routledge, 2002).
11. Atherton, M. *Gambling* (London:Hachette, 2007).
12. *The Guardian*. Pakistan Embroiled in No-Ball Betting Scandal against England. 29 Aug 2010. Available from: http://www.guardian.co.uk/sport/2010/aug/29/pakistan-cricket-betting-allegations.
13. Gamble Aware :: Gambling Facts and Figures. 2012. Available from: http://www.gambleaware.co.uk/gambling-facts-and-figures.
14. Dubins, L.E., Savage, L.J. *How To Gamble If You Must: Inequalities for Stochastic Processes* (New York: McGraw-Hill, 1965).
15. The WLA Security Control Standard (certification documents). 2012. Available from: https://www.world-lotteries.org/cms/index.php?option=com_content&view=article&id=4374%3Athe-wla-security-control-standard-2012-wla-scs2012&catid=106%3Asecurity-control-standard-scs&Itemid=100177&lang=en.
16. *Daily Mirror*. Gambler Wins £585k for 86p Stake on 19-Match Accumulator. 2011. Available from: http://www.mirrorfootball.co.uk/news/Gambler-wins-585k-for-86p-stake-on-19-match-accumulator-thanks-to-Glen-Johnson-87th-minute-Liverpool-winner-v-Chelsea-article833317.html.
17. DSM-IV Criteria: Pathological Gambling. 2000. Available from: http://www.problemgambling.ca/EN/ResourcesForProfessionals/Pages/DSMIVCriteriaPathologicalGambling.aspx.

18. Central and North West London NHS Foundation Trust :: Gambling Treatment Centre London. 2012. Available from: http://www.cnwl.nhs.uk/gambling.html.

Chapter 13: Average Risks

1. BBC. Figures show 'Mr and Mrs Average'. 13 Oct 2010. Available from: http://www.bbc.co.uk/news/uk-11534042.
2. Office for National Statistics. 2011 Annual Survey of Hours and Earnings (SOC 2000). 2011. Available from: http://www.ons.gov.uk/ons/rel/ashe/annual-survey-of-hours-and-earnings/ashe-results-2011/ashe-statistical-bulletin-2011.html.
3. Gould, S.J. Median Is Not the Message. Available from: http://people.umass.edu/biep540w/pdf/Stephen%20Jay%20Gould.pdf.
4. Dilnot, A., Blastland, M. *The Tiger That Isn't: Seeing Through a World of Numbers* (London:Profile Books, 2010).
5. Savage, S.L.. *The Flaw of Averages: Why We Underestimate Risk in the Face of Uncertainty* (Chichester: John Wiley, 2009).

Chapter 14: Chance

1. Dostoyevsky, F. *Notes from the Underground and The Double* (Harmondsworth: Penguin Books, 1972).
2. The Information Philosopher. Chrysippus. 2012. Available from: http://www.informationphilosopher.com/solutions/philosophers/chrysippus/.
3. Oppenheimer, J.R. *Science and the Common Understanding* (New York: Simon and Schuster, 1954).
4. Adams, D. *The Hitchhiker's Guide to the Galaxy* (New York: Random House, 1997).
5. Rich, M.D. *A Million Random Digits with 100,000 Normal Deviates* (Santa Monica, Rand Corporation, 2001).
6. See Extreme Fire Behavior, *Atlantic Magazine* (September 2012).

Chapter 15: Transport

1. *The Guardian*. Life Sentence for Train Murder of Student. 10 Nov 2006. Available from: http://www.guardian.co.uk/uk/2006/nov/10/ukcrime.
2. Currie, G., Delbosc, A., Mahmoud, S. Perceptions and Realities of Personal Safety on Public Transport for Young People in Melbourne. Conference paper delivered at the Australasian Transport Research Forum held in Canberra, Australia 2010.

3. Rail Safety and Standards Board. Annual Safety Performance Report 2010/11. 2011. Available from: http://www.rssb.co.uk/SPR/REPORTS/Documents/ASPR%202010-11%20Final.pdf.

4. Evans, A. Fatal Train Accidents on Britain's Main Line Railways: End of 2010 Analysis. Centre for Transport Studies, Imperial College London. Available from: http://www.cts.cv.ic.ac.uk/documents/publications/iccts01391.pdf.

5. BBC News. Guard Jailed over Rail Death Fall. 15 Nov 2012. Available from: http://www.bbc.co.uk/news/uk-england-merseyside-20339630.

6. See note 3 above.

7. Wolff, J. Risk, Fear, Blame, Shame and the Regulation of Public Safety. *Economics and Philosophy*. 2006;22(03):409–27.

8. Gigerenzer, G. Out of the Frying Pan into the Fire: Behavioral Reactions to Terrorist Attacks. *Risk Analysis*. 2006 Apr 1;26(2):347–51.

9. Hoorens, V. Self-Enhancement and Superiority Biases in Social Comparison. *European Review of Social Psychology*. 1993;4(1):113–39; McCormick, I.A., Walkey, F.H., Green, D.E. Comparative Perceptions of Driver Ability – A Confirmation and Expansion. *Accid Anal Prev*. 1986 Jun;18(3):205–8.

10. International Traffic Safety Data and Analysis Group. IRTAD Road Safety Annual Report. 2011. Available from: http://internationaltransportforum.org/irtadpublic/index.html.

11. World Health Organisation. Global Status Report on Road Safety. 2009. Available from: http://www.who.int/violence_injury_prevention/road_safety_status/2009/en/.

12. FIA Foundation. The Missing Link: Road Traffic Injuries and the Millennium Development Goals. 2010. Available from: http://www.fiafoundation.org/publications/Pages/PublicationHome.aspx.

13. See note 11 above.

14. Smeed, R.J. Some Statistical Aspects of Road Safety Research. *Journal of the Royal Statistical Society*, Series A (General). 1949;112(1):1.

15. Oakes, M., Bor, R. The Psychology of Fear of Flying (Part I): A Critical Evaluation of Current Perspectives on the Nature, Prevalence and Etiology of Fear of Flying. *Travel Med Infect Dis*. 2010 Nov;8(6):327–38.

16. British Airways. Flying with Confidence – Fear of Flying Course from British Airways [Internet]. Available from: http://flyingwithconfidence.com/.

17. Plane Crash Info. Available from: http://planecrashinfo.com/index.html.

18. Ibid.

19. Ibid.

20. NTSB National Transportation Safety Board. NTSB – Aviation Statistical Reports. Available from: http://www.ntsb.gov/data/aviation_stats.html.

21. Department for Transport. Aviation – Statistics. Available from: https://www.gov.uk/government/organisations/department-for-transport/series/aviation-statistics.
22. NTSB National Transportation Safety Board. Preliminary Monthly Summary. 2012. Available from: http://www.ntsb.gov/data/monthly/curr_mo.TXT.

Chapter 16: Extreme Sports

1. The Best Wing Suit /Skydive from YouTube, part 1. 2008. Available from: http://www.youtube.com/watch?v=5N9t5qOSzCU&feature=youtube_gdata_player.
2. *Daily Mail*. Last One on the Ground Is a Rotten Egg! Spectacular Photos of Daredevils Diving in Base-Jumping Race. Available from: http://www.dailymail.co.uk/news/article-2164332/World-Base-Race-2012-Spectacular-photos-daredevils-diving-base-jumping-race.html.
3. Clark, R.W. *The Victorian Mountaineers* (London: Batsford, 1953).
4. Windsor, J.S., Firth, P.G., Grocott, M.P., Rodway, G.W., Montgomery, H.E. Mountain Mortality: A Review of Deaths that Occur during Recreational Activities in the Mountains. *Postgraduate Medical Journal*. 2009 Jun 1;85(1004):316–21.
5. Pollard, A., Clarke, C. Deaths during Mountaineering at Extreme Altitude. *The Lancet*. 1988 Jun;331(8597):1277.
6. Wikipedia. Franz Reichelt. Available from: http://en.wikipedia.org/wiki/Franz_Reichelt. The terrifying film of Reichelt's jump is on YouTube: http://www.youtube.com/watch?v=BepyTSzueno.
7. United States Parachute Association. Skydiving Safety. Available from: http://www.uspa.org/AboutSkydiving/SkydivingSafety/tabid/526/Default.aspx.
8. Soreide, K., Ellingsen, C.L., Knutson, V. How Dangerous Is BASE Jumping? An Analysis of Adverse Events in 20,850 Jumps From the Kjerag Massif, Norway. *The Journal of Trauma: Injury, Infection, and Critical Care*. 2007 May;62(5):1113–7.
9. British Sub-Aqua Club. UK Diving Fatalities Review. Available from: http://www.bsac.com/page.asp?section=3780&sectionTitle=UK+Diving+Fatalities+Review.
10. Redelmeier, D.A., Greenwald, J.A. Competing Risks of Mortality with Marathons: Retrospective Analysis. *BMJ*. 2007 Dec 22;335(7633):1275–7.
11. Kipps, C., Sharma, S., Pedoe, D.T. The Incidence of Exercise-Associated Hyponatraemia in the London Marathon. *British Journal of Sports Medicine*. 2011 Jan 1;45(1):14–19.

12. Royal Society for the Prevention of Accidents. Home and Leisure Accident Statistics: RoSPA : HASS and LASS. Available from: http://www. hassandlass.org.uk/query/index.htm.

13. See note 10 above.

14. Bennett, P., Calman, K., Curtis, S., Fischbacher-Smith, D. *Risk Communication and Public Health* (Oxford: Oxford University Press; 2009).

Chapter 17: Lifestyle

1. *Daily Express.* Less Meat, More Veg is the Secret for Longer Life. 2012. Available from: http://www.express.co.uk/posts/view/307781; Pan, A., Sun, Q., Bernstein, A.M., Schulze, M.B., Manson, J.E., Stampfer, M.J., et al. Red Meat Consumption and Mortality: Results From 2 Prospective Cohort Studies. *Arch Intern Med.* 2012 Apr 9;172(7):555–63.

2. Partington, A. *The Oxford Dictionary of Quotations* (Oxford: Oxford University Press, 1996).

3. Shaw, M., Mitchell, R., Dorling, D. Time for a Smoke? One Cigarette Reduces Your Life by 11 Minutes. *BMJ.* 2000 Jan 1;320(7226):53.

4. Doll R. Mortality in Relation to Smoking: 50 Years' Observations on Male British Doctors. *BMJ.* 2004 Jun 26;328:1519–20.

5. Spiegelhalter, D. Using Speed of Ageing and 'Microlives' to Communicate the Effects of Lifetime Habits and Environment. *BMJ.* 17 Dec 2012;345(dec14 14):e8223–e8223.

6. Prospective Studies Collaboration. Body-Mass Index and Cause-Specific Mortality in 900,000 Adults: Collaborative Analyses of 57 Prospective Studies. *The Lancet.* 2009 Mar;373:1083–96.

7. Khaw K.-T., Wareham N., Bingham S., Welch A., Luben R., Day N. Combined Impact of Health Behaviours and Mortality in Men and Women: The EPIC-Norfolk Prospective Population Study. *PLoS Med.* 2008;5(1):e12

8. NHS Information Centre. Statistics on Obesity, Physical Activity and Diet: England. 2012. Available from: http://www.hscic.gov.uk/searchcatalogue?pr oductid=4787&topics=2%2fPublic+health%2fLifestyle%2fPhysical+activit y&sort=Relevance&size=10&page=1.

9. Woodcock, J., Franco, O.H., Orsini, N., Roberts, I. Non-Vigorous Physical Activity and All-Cause Mortality: Systematic Review and Meta-Analysis of Cohort Studies. *Int. J. Epidemiol.* 2011 Feb;40(1):121–38.

10. Byberg, L., Melhus, H., Gedeborg, R., Sundström, J., Ahlbom, A., Zethelius, B., et al. Total Mortality after Changes in Leisure Time Physical Activity in 50 Year Old Men: 35 Year Follow-Up of Population Based Cohort. *BMJ.* 2009;338:b688.

11. Ben Goldacre. Vitamin Pills Can Lead You To Take Health Risks. *The Guardian*. 2011. Available from: http://www.guardian.co.uk/commentisfree/2011/aug/26/bad-science-vitamin-pills-lead-you-to-take-risks.

12. Shaw, K.A., Gennat, H.C., O'Rourke, P., Del Mar, C. Exercise for Overweight or Obesity. Cochrane Database of Systematic Reviews. John Wiley & Sons, Ltd. 1996. Available from: http://onlinelibrary.wiley.com/doi/10.1002/14651858.CD003817.pub3/abstract.

Chapter 18: Health and Safety

1. Health and Safety Executive. Myth: You Can't Throw Out Sweets at Pantos. 2009. Available from: http://www.hse.gov.uk/myth/dec09.htm.

2. *Daily Telegraph*. Health and Safety Fears Are 'Taking the Joy out of Playtime'. Telegraph.co.uk. 1 Jul 2011. Available from: http://www.telegraph.co.uk/education/educationnews/8612145/Health-and-safety-fears-are-taking-the-joy-out-of-playtime.html.

3. Adams, J. Risk in a Hypermobile World. 2012. Available from: http://www.john-adams.co.uk/.

4. *Hazards*. Hazards 117, January–March 2012. Available from: http://www.hazards.org/haz117/index.htm.

5. Health and Safety Executive. HSE Statistics: Historical Picture. 2000. Available from: http://www.hse.gov.uk/statistics/history/index.htm.

6. Health and Safety Executive. Self-Reported Work-Related Illness and Workplace Injuries. 2008. Available from: http://www.hse.gov.uk/statistics/lfs/index.htm.

7. Health and Safety Executive. Workplace Fatalities and Injuries Statistics in the EU [Internet]. 2008. Available from: http://www.hse.gov.uk/statistics/european/index.htm.

8. Bureau of Labor Statistics. Census of Fatal Occupational Injuries (CFOI) – Current and Revised Data. Available from: http://www.bls.gov/iif/oshcfoi1.htm.

9. International Labour Organisation. Global Workplace Deaths Vastly Under-Reported, Says ILO [Internet]. Available from: http://www.ilo.org/global/about-the-ilo/press-and-media-centre/news/WCMS_005176/lang--en/index.htm.

10. International Labour Organisation. XIX World Congress on Safety and Health at Work – ILO Introductory Report: Global Trends and Challenges on Occupational Safety and Health. Available from: http://www.ilo.org/safework/info/publications/WCMS_162662/lang--en/index.htm.

11. Asian Development Bank. People's Republic of China: Coal Mine Safety Study. Part II: Review and Analysis of International Experience.

2007. Available from: http://www.adb.org/Documents/Reports/
Consultant/39657-PRC/39657-02-PRC-TACR.pdf.

12. Department of Energy and Climate Change. Coal Mining Technologies and
Production Statistics – The Coal Authority. 2012. Available from: http://
coal.decc.gov.uk/en/coal/cms/publications/mining/mining.aspx.

13. *Hazards*. Safety in the Pits – *Hazards* issue 116, October–December 2011.
Available from: http://www.hazards.org/deadlybusiness/deadlymines.htm.

14. Ibid.

15. Tu, J. Coal Mining Safety: China's Achilles' Heel. *China Security*.
2007;3:36–53.

16. See note 11 above.

17. See note 13 above.

18. Roberts, S.E. Britain's Most Hazardous Occupation: Commercial Fishing.
Accident Analysis & Prevention. 2010 Jan;42(1):44–9.

19. h2g2. The London Beer Flood of 1814. 2012. Available from: http://h2g2.
com/dna/h2g2/A42129876.

20. Wikipedia. Boston Molasses Disaster. 2012. Available from: http://
en.wikipedia.org/wiki/Boston_Molasses_Disaster.

21. Wikipedia. Bhopal Disaster. 2012. Available from: http://en.wikipedia.org/
wiki/Bhopal_disaster.

22. See note 5 above.

23. Albin, M., Horstmann, V., Jakobsson, K., Welinder, H. Survival in Cohorts
of Asbestos Cement Workers and Controls. *Occup Environ Med*. 1996 Feb
1;53(2):87–93.

24. Miller, B.G., Jacobsen, M. Dust Exposure, Pneumoconiosis, and Mortality of
Coalminers. *Br. J. Ind. Med*. 1985 Nov 1;42(11):723–33.

25. Health Safety Executive R. Reducing Risks, Protecting People. HSE's
Decision-Making Process. 2001. Available from: http://www.hse.gov.uk/risk/
theory/r2p2.htm.

26. Bird, S., Fairweather, C. Revent Military Fatalities in Afghanistan by Cause
and Nationality: Period 15, 5 September 2011 to 22 January 2012. Available
from: http://www.mrc-bsu.cam.ac.uk/Publications/PDFs/PERIOD_15_
fatalities_in_Afghanistan_by_cause_and_nationality.pdf.

Chapter 19: Radiation

1. Adams J. Risk in a Hypermobile World. 2012. Available from: http://www.
john-adams.co.uk/.

2. Slovic P. Perception of Risk. *Science*. 1987 Apr 17;236(4799):280–5.

3. Mehta, P., Smith-Bindman, R. Airport Full-Body Screening: What Is the
Risk? *Arch Intern Med*. 2011 Jun 27;171(12):1112–5.

4. Health ProtectionAgency. Dose Comparisons for Ionising Radiation. Available from: http://www.hpa.org.uk/Topics/Radiation/ UnderstandingRadiation/UnderstandingRadiationTopics/ DoseComparisonsForIonisingRadiation/. World Health Organisation (2012) Preliminary Dose Estimation from the nuclear accident after the 2011 Great East Japan Earthquake and Tsunami. http://www.who.int/ionizing_ radiation/pub_meet/fukushima_dose_assessment/en/index.html.

5. National Academy of Sciences. Health Effects of Radiation: Findings of the Radiation Effects Research Foundation. 2003. Available from: http://dels-old.nas.edu/dels/rpt_briefs/rerf_final.pdf.

6. United Nations Scientific Committee on the Effects of Atomic Radiation. UNSCEAR Assessments of the Chernobyl Accident. 2012. Available from: http://www.unscear.org/unscear/en/chernobyl.html.

7. Little, M.P., Hoel, D.G., Molitor, J., Boice, J.D., Wakeford, R., Muirhead, C.R. New Models for Evaluation of Radiation-Induced Lifetime Cancer Risk and its Uncertainty Employed in the UNSCEAR 2006 Report. *Radiat. Res.* 2008 Jun;169(6):660–76.

8. Berrington de Gonzalez, A., Mahesh, M., Kim, K.-P., Bhargavan, M., Lewis, R., Mettler, F., et al. Projected Cancer Risks from Computed Tomographic Scans Performed in the United States in 2007. *Arch Intern Med.* 2009 Dec 14;169(22):2071–7.

9. EU Business News. After Japan 'Apocalypse', EU Agrees Nuclear 'Stress Tests'. Available from: http://www.eubusiness.com/news-eu/ japan-quake-nuclear.93d.

Chapter 20: Space

1. BBC News. How Often Do Plane Stowaways Fall from the Sky? 13 Sep 2012. Available from: http://www.bbc.co.uk/news/magazine-19562101.

2. Sagan C. *Pale Blue Dot: A Vision of the Human Future in Space* (New York, Random House, 1994).

3. *The Guardian*. Comette Family Home Damaged by Egg-Sized Meteorite. 2011. Available from: http://www.guardian.co.uk/world/2011/oct/10/ comette-family-home-damaged-meteorite.

4. Woo G. *Calculating Catastrophe* (London, Imperial College Press, 2011).

5. Risk Management Solutions. Comet and Asteroid Risk: An Analysis of the 1908 Tunguska Event. 2009. Available from: www.rms.com/ publications/1908_tunguska_event.pdf.

6. National Research Council. *Defending Planet Earth:Near-Earth Object Surveys and Hazard Mitigation Strategies* (Washington, D.C.: The National Academies Press, 2010).

7. Boslough, M.B.E., Crawford, D.A. Low-Altitude Airbursts and the Impact Threat. *International Journal of Impact Engineering*. 2008 Dec;35(12):1441–8.

8. Ward, S.N., Asphaug, E. Asteroid Impact Tsunami: A Probabilistic Hazard Assessment. *Icarus*. 2000 May;145(1):64–78.

9. See note 4 above.

10. NASA. Near-Earth Object Program. Available from: http://neo.jpl.nasa.gov/index.html.

11. NASA NEO Program. The Torino Impact Hazard Scale. Available from: http://neo.jpl.nasa.gov/torino_scale.html.

12. NASA NEO Program. Predicting Apophis' Earth Encounters in 2029 and 2036. Available from: http://neo.jpl.nasa.gov/apophis/.

13. Discovery News. Hayabusa Asteroid Probe Awarded World Record : Discovery News. Available from: http://news.discovery.com/space/hayabusa-asteroid-probe-gets-guinness-world-record-110620.html.

Chapter 21: Unemployment

1. Taleb, N. *The Black Swan: The Impact of the Highly Improbable* (New York: Random House, 2007).

2. Slovic, P. *The Perception of Risk* (London: Routledge, 2000).

3. Office for National Statistics. Labour Market Flows, November 2011 (Experimental Statistics). 2011. Available from: http://www.ons.gov.uk/ons/rel/lms/labour-market-statistics/november-2011/art-labour-market-flows--july-to-september-2011.html.

4. Thomas, K. *The Oxford Book of Work* (Oxford: Oxford University Press, 1999).

5. Trades Union Congress. The Costs of Unemployment. 2010. Available from: http://www.tuc.org.uk/extras/costsofunemployment.pdf.

6. Ruhm, C.J. Are Recessions Good for Your Health? *The Quarterly Journal of Economics*. 2000 May 1;115(2):617–50.

7. National Institute for Health and Clinical Excellence. Worklessness and Health: What Do We Know about the Causal Relationship? 2005. Available from: http://www.nice.org.uk/niceMedia/documents/worklessness_health.pdf.

8. Roelfs, D.J., Shor, E., Davidson, K.W., Schwartz, J.E. Losing Life and Livelihood: A Systematic Review and Meta-Analysis of Unemployment and All-Cause Mortality. *Soc. Sci. Med.* 2011 Mar;72(6):840–54.

9. Clemens, T., Boyle, P., Popham, F. Unemployment, Mortality and the Problem of Health-Related Selection: Evidence from the Scottish and England & Wales (ONS) Longitudinal Studies. *Health Stat. Q.* 2009;(43):7–13.

10. Gregg, P., Tominey, E. The Wage Scar from Youth Unemployment. Department of Economics, University of Bristol; 2004. Report No.: 04/097. Available from: http://ideas.repec.org/p/bri/cmpowp/04-097.html.

11. Bell, D.N.F., Blanchflower, D.G. Youth Unemployment: Déjà Vu? Institute for the Study of Labor (IZA); 2010. Report No.: 4705. Available from: http://ideas.repec.org/p/iza/izadps/dp4705.html.

Chapter 22: Crime

1. *The Guardian*. Assaulted Pensioner Emma Winnall Dies of Her Injuries. 2012. Available from: http://www.guardian.co.uk/uk/2012/may/29/assaulted-pensioner-emma-winnall-dies.

2. Home Office. Crime in England and Wales: Quarterly Update to September 2011. Available from: https://www.gov.uk/government/publications/crime-in-england-and-wales-quarterly-update-to-september-2011.

3. BBC News. Whispering Game. 16 Feb 2006. Available from: http://news.bbc.co.uk/1/hi/magazine/4719364.stm.

4. Kahneman, D. *Thinking, Fast and Slow* (New York: Farrar, Straus and Giroux, 2011).

5. Slovic P. If I Look at the Mass I Will Never Act: Psychic Numbing and Genocide. In: Roeser, S., ed. *Emotions and Risky Technologies* (Dordrect: Springer, 2010), p. 37–59. Available from: http://link.springer.com/chapter/10.1007/978-90-481-8647-1_3.

6. Fagerlin, A., Wang, C., Ubel, P.A. Reducing the Influence of Anecdotal Reasoning on People's Health Care Decisions: Is a Picture Worth a Thousand Statistics? *Med. Decis. Making*. 2005 Aug;25(4):398–405.

7. Wood, James. *How Fiction Works* (London: Cape, 2008).

8. Home Office. Homicides, Firearm Offences and Intimate Violence 2010/11: Supplementary Vol. 2 to Crime in England and Wales. 2012. Available from: https://www.gov.uk/government/publications/homicides-firearm-offences-and-intimate-violence-2010-to-2011-supplementary-volume-2-to-crime-in-england-and-wales-2010-to-2011.

9. BBC News. Brown Pledge to Tackle Stabbings. 11 Jul 2008. Available from: http://news.bbc.co.uk/1/hi/uk/7502569.stm.

10. Spiegelhalter, D., Barnett, A. London Murders: A Predictable Pattern? *Significance*. 2009;6(1):5–8.

11. See note 2 above.

12. Ibid.

13. Home Office. Crime in England and Wales 2006/2007. Available from: http://webarchive.nationalarchives.gov.uk/20110220105210/http:/rds.homeoffice.gov.uk/rds/crimeew0607.html.

Chapter 23: Surgery

1. Longer, Healthier, Happier? Human Needs, Human Values and Science. Sense about Science annual lecture. 2007 Available from: http://www. senseaboutscience.org/pages/annual-lecture-2007.html.

2. Isaacs, D., Fitzgerald, D. Seven Alternatives to Evidence Based Medicine. *BMJ*. 1999 Dec 18;319(7225):1618.

3. Guyatt, G.C.J. Evidence-Based Medicine: A New Approach to Teaching the Practice of Medicine. *JAMA*. 1992 Nov 4;268(17):2420–5.

4. Ioannidis, J.P.A. Why Most Published Research Findings Are False. *PLoS Med*. 2005;2(8):e124.

5. Gawande, A. *Complications: A Surgeon's Notes on an Imperfect Science* (New York: Henry Holt, 2002).

6. Gross, C.G. *A Hole in the Head: More Tales in the History of Neuroscience* (Cambridge, MA:MIT Press, 2009).

7. Weiser, T.G., Regenbogen, S.E., Thompson, K.D., Haynes, A.B., Lipsitz, S.R., Berry, W.R., et al. An Estimation of the Global Volume of Surgery: A Modelling Strategy Based on Available Data. *Lancet*. 2008 Jul 12;372(9633):139–44.

8. Royal College of Anaesthetists. Risks Associated with your Anaesthetic Section 14: Death or Brain Damage. 2009. Available from: http://www.rcoa. ac.uk/node/2066.

9. Nightingale, F. *Notes on Hospitals* (London: Longman, Green, Longman, Roberts and Green, 1863).

10. Codman, E.A. *A Study in Hospital Efficiency: As Demonstrated by the Case Report of the First Five Years of a Private Hospital* (Boston: T. Todd, 1918).

11. Ferguson, T.B. Jr, Hammill, B.G., Peterson, E.D., DeLong, E.R., Grover, F.L. A Decade of Change – Risk Profiles and Outcomes for Isolated Coronary Artery Bypass Grafting Procedures, 1990-1999: A Report from the STS National Database Committee and the Duke Clinical Research Institute. Society of Thoracic Surgeons. *Ann. Thorac. Surg.* 2002 Feb;73(2):480–489 (discussion 489–90).

12. Care Quality Commission. Survival Rates – Heart Surgery in United Kingdom 2008–9. Available from: http://heartsurgery.cqc.org.uk/survival. aspx.

13. New York State Department of Health. Cardiovascular Disease Data and Statistics. 2012. Available from: http://www.health.ny.gov/statistics/diseases/ cardiovascular/.

14. Campbell, M.J., Jacques, R.M., Fotheringham, J., Maheswaran, R., Nicholl, J. Developing a Summary Hospital Mortality Index: Retrospective Analysis in English Hospitals over Five Years. *BMJ*. 2012 Mar 1;344(mar01 1):e1001–e1001.

15. Hawkes N. Patient Coding and the Ratings Game. *BMJ*. 2010 Apr 25;340(apr23 2):c2153–c2153.

Chapter 24: Screening

1. Cancer Research UK. Breast Screening: Accuracy of Mammography. 2012. Available from: http://www.cancerresearchuk.org/cancer-info/cancerstats/types/breast/screening/Other-Issues/#Accuracy.
2. McCartney, M. *The Patient Paradox* (London: Pinter & Martin, 2012).
3. Mehta, P., Smith-Bindman, R. Airport Full-Body Screening: What Is the Risk? *Arch Intern Med*. 2011 Jun 27;171(12):1112–5.
4. NHS Breast Screening Programme. Screening for Breast Cancer in England: Past and Future. 2012. Available from: http://www.cancerscreening.nhs.uk/breastscreen/publications/nhsbsp61.html.
5. Yaffe, M.J., Mainprize, J.G. Risk of Radiation-Induced Breast Cancer from Mammographic Screening. *Radiology*. 2011 Jan;258(1):98–105.
6. Welch, H.G., Schwartz, L.M., Woloshin, S. *Overdiagnosed: Making People Sick in the Pursuit of Health* (Boston, MA: Beacon Press, 2011).
7. Cancer Research UK. Breast Screening Review. 2012. Available from: http://www.cancerresearchuk.org/cancer-info/publicpolicy/ourpolicypositions/symptom_Awareness/cancer_screening/breast-screening-review/breast-screening-review.
8. Ibid.
9. Ablin, R.J. The Great Prostate Mistake. *The New York Times* (10 Mar 2010). Available from: http://www.nytimes.com/2010/03/10/opinion/10Ablin.html.
10. House of Lords. Lord Andrew Lloyd-Webber Has Called for All Men over the Age of 50 to Have a Test for Prostate Cancer. BBC. 19 Jul 2010. Available from: http://news.bbc.co.uk/democracylive/hi/house_of_lords/newsid_8822000/8822506.stm.
11. *Daily Mail*. Andrew Lloyd Webber Reveals Prostate Cancer Battle Has Left Him Impotent. 2011. Available from: http://www.dailymail.co.uk/tvshowbiz/article-1371379/Andrew-Lloyd-Webber-reveals-prostate-cancer-battle-left-impotent.html.
12. Vernooij, M.W., Ikram, M.A., Tanghe, H.L., Vincent, A.J.P.E., Hofman, A., Krestin, G.P., et al. Incidental Findings on Brain MRI in the General Population. *New England Journal of Medicine*. 2007 Nov 1;357(18):1821–8.
13. Cancer Research UK. Prostate Cancer – UK Incidence Statistics. 2011. Available from: http://info.cancerresearchuk.org/cancerstats/types/prostate/incidence/.
14. See note 6 above.

15. Andriole, G.L., Crawford, E.D., Grubb, R.L., Buys, S.S., Chia, D., Church, T.R., et al. Prostate Cancer Screening in the Randomized Prostate, Lung, Colorectal, and Ovarian Cancer Screening Trial: Mortality Results after 13 Years of Follow-Up. *J. Natl. Cancer Inst.* 2012 Jan 18;104(2):125–32.

16. Schröder, F.H., Hugosson, J., Roobol, M.J., Tammela, T.L.J., Ciatto, S., Nelen, V., et al. Prostate-Cancer Mortality at 11 Years of Follow-Up.. *New England Journal of Medicine.* 2012 Mar 15;366(11):981–90.

17. Welch, H.G., Frankel, B.A. Likelihood That a Woman With Screen-Detected Breast Cancer Has Had Her 'Life Saved' by That Screening. *Arch Intern Med.* 2011 Dec 12;171(22):2043–6.

18. 23andMe. Genetic Testing for Health, Disease & Ancestry; DNA Test. Available from: https://www.23andme.com/.

Chapter 25: Money

1. Fowler, S. Workhouse: The People, the Places, the Life behind Doors. National Archives; 2007.

2. Booth, Charles. The Aged Poor in England and Wales (London, 1894). Available from: http://archive.org/stream/agedpoorinenglaoobootgoog/agedpoorinenglaoobootgoog_djvu.txt.

3. Crowther, M.A. *The Workhouse System, 1834–1929: The History of an English Social Institution* (London: Routledge, 1983).

4. Orwell, G. *Down and Out in Paris and London: A Novel* (Harcourt, Brace & World, 1961).

5. Households Below Average Income. Department of Work and Pensions, June 2012. Available from: http://research.dwp.gov.uk/asd/index.php?page=hbai.

6. Ibid.

7. English Longitudinal Study of Ageing. Financial Circumstances, Health and Well-Being of the Older Population in England. 2010. Available from: http://www.ifs.org.uk/elsa/report10/elsa_w4-1.pdf.

8. Department of Health. Fairer Care Funding. The Report of the Commission on Funding of Care and Support. 2011. Available from: http://www.dilnotcommission.dh.gov.uk/our-report/.

Chapter 26: The End

1. Halley, E. An Estimate of the Degrees of the Mortality of Mankind, Drawn from Curious Tables of the Births and Funerals at the City of Breslaw; With an Attempt to Ascertain the Price of Annuities upon Lives. Royal Society of London; 1753. Available from: http://archive.org/details/philtrans05474358.

2. Human Mortality Database. Human Mortality Database. Available from: http://www.mortality.org/.

3. Office for National Statistics. Life Expectancy at Birth and at Age 65 by Local Areas in the United Kingdom. 2011. Available from: http://www.ons.gov.uk/ons/rel/subnational-health4/life-expec-at-birth-age-65/2004-06-to-2008-10/index.html.

4. Understanding Uncertainty. Survival Worldwide. 2012. Available from: http://understandinguncertainty.org/node/272.

5. World Health Organisation. World Health Statistics 2011. Available from: http://www.who.int/gho/publications/world_health_statistics/2011/en/index.htm.

6. Human Life-Table Database. 2012. Available from: http://www.lifetable.de/.

7. Office for National Statistics. Period and Cohort Life Expectancy Tables. 2011. Available from: http://www.ons.gov.uk/ons/rel/lifetables/period-and-cohort-life-expectancy-tables/2010-based/index.html.

8. Office for National Statistics. What Are the Chances of Surviving to Age 100? 2012. Available from: http://www.ons.gov.uk/ons/rel/lifetables/historic-and-projected-mortality-data-from-the-uk-life-tables/2010-based/rpt-surviving-to-100.html.

9. United Nations. Global Issues at the United Nations. Available from: http://www.un.org/en/globalissues/ageing/.

10. Collerton, J., Davies, K., Jagger, C., Kingston, A., Bond, J., Eccles, M.P., et al. Health and Disease in 85 Year Olds: Baseline Findings from the Newcastle 85+ Cohort Study. *BMJ* [Internet]. 2009;339. Available from: http://www.ncbi.nlm.nih.gov/pmc/articles/PMC2797051/.

11. Office for National Statistics. Pension Trends, Chapter 2: Population Change. 2012. Available from: http://www.ons.gov.uk/ons/rel/pensions/pension-trends/chapter-2--population-change--2012-edition-/index.html.

12. Knapp, M., Prince, M. Dementia UK 2007. Available from: http://alzheimers.org.uk/site/scripts/download_info.php?fileID=2.

13. Office for National Statistics. Pension Trends, Chapter 3: Life Expectancy and Healthy Ageing. 2012. Available from: http://www.ons.gov.uk/ons/rel/pensions/pension-trends/chapter-3--life-expectancy-and-healthy-ageing--2012-edition-/index.html.

14. British Parachute Association. How Safe? 2012. Available from: http://www.bpa.org.uk/staysafe/how-safe/.

Chapter 27: Judgement Day

1. De Finetti, B. *Theory of Probability* (London: John Wiley, 1974).

ACKNOWLEDGEMENTS

Andrew Franklin of Profile Books first suggested that we write together, an idea for which we're hugely grateful and which lacked only the detail of something to write about. For that detail, we owe thanks for two professional lifetimes of influences and opportunities and innumerable (even for us) friends and colleagues who helped us along the way. The Profile team has been a joy to work with, as ever. Jonny Pegg has been all you would want from an agent – in fact a double agent – a cogent critic and an affable, encouraging friend. Andrew Dilnot gave invaluable advice at an early stage. Rich Knight and Chris Vince read some early material and said the idea wasn't unhinged, which helped. Katey Adderley and Caitlin Harris cheered MB along and were a mine of psychological insight about risk, while Joe Harris gave him his most important lesson in how to live with it. Edgar and Kieran supplied the outline of the story about penguins, years ago and from life. Fiona revealed what you can do with a Fray Bentos Pie – from life – and we shamelessly plundered the thoughts about danger of any other friends foolish enough to talk about it. Kate Bull gave unlimited encouragement and advice, and even liked it, which was the main thing. Mike Pearson has provided endless support and inspiration. David Harding's generosity gave DS the opportunity to spend his time writing this sort of stuff. Thanks to them all.

INDEX